AF389454

Optical Signal Processing Technologies for Communication, Computing, and Sensing Applications

Optical Signal Processing Technologies for Communication, Computing, and Sensing Applications

Editors

Jiangbing Du
Yang Yue
Jian Zhao
Yan-ge Liu

MDPI • Basel • Beijing • Wuhan • Barcelona • Belgrade • Manchester • Tokyo • Cluj • Tianjin

Editors
Jiangbing Du
Shanghai Jiao Tong University,
Shanghai, China

Yang Yue
Xi'an Jiaotong University,
Xi'an, China

Jian Zhao
Tianjin University,
Tianjin, China

Yan-ge Liu
Nankai University,
Tianjin, China

Editorial Office
MDPI
St. Alban-Anlage 66
4052 Basel, Switzerland

This is a reprint of articles from the Special Issue published online in the open access journal *Sensors* (ISSN 1424-8220) (available at: https://www.mdpi.com/journal/sensors/special_issues/OCS).

For citation purposes, cite each article independently as indicated on the article page online and as indicated below:

LastName, A.A.; LastName, B.B.; LastName, C.C. Article Title. *Journal Name* **Year**, *Volume Number*, Page Range.

ISBN 978-3-0365-7152-2 (Hbk)
ISBN 978-3-0365-7153-9 (PDF)

Contents

Jiangbing Du, Yang Yue, Jian Zhao and Yange Liu
Editorial: Special Issue "Optical Signal Processing Technologies for Communication,
Computing, and Sensing Applications"
Reprinted from: *Sensors* **2023**, *23*, 2606, doi:10.3390/s23052606 . . . . **1**

Yuzhu Zhu, Jiangbing Du, Weihong Shen and Zuyuan He
Nonlinear Distortion by Stimulated Brillouin Scattering in Kramers-Kronig Receiver Based
Optical Transmission
Reprinted from: *Sensors* **2022**, *22*, 7287, doi:10.3390/s22197287 . . . . **3**

Jian Yang, Yingning Wang, Yuxi Fang, Wenpu Geng, Wenqian Zhao, Changjing Bao, et al.
Over-Two-Octave Supercontinuum Generation of Light-Carrying Orbital Angular Momentum
in Germania-Doped Ring-Core Fiber
Reprinted from: *Sensors* **2022**, *22*, 6699, doi:10.3390/s22176699 . . . . **11**

Paweł Rosa, Giuseppe Rizzelli Martella and Mingming Tan
Bandwidth Extension in a Mid-Link Optical Phase Conjugation
Reprinted from: *Sensors* **2022**, *22*, 6385, doi:10.3390/s22176385 . . . . **23**

**Carmen Álvarez-Roa, María Álvarez-Roa, Francisco J. Martín-Vega, Miguel
Castillo-Vázquez, Thiago Raddo and Antonio Jurado-Navas**
Performance Analysis of a Vertical FSO Link with Energy Harvesting Strategy
Reprinted from: *Sensors* **2022**, *22*, 5684, doi:10.3390/s22155684 . . . . **35**

Luis M. Torres, Francisco J. Cañete and Luis Díez
Matched Filtering for MIMO Coherent Optical Communications with Mode-Dependent Loss
Channels
Reprinted from: *Sensors* **2022**, *22*, 798, doi:10.3390/s22030798 . . . . **57**

Yanliang Duan, Xinhua Yu, Lirong Mei and Weiping Cao
Low-Complexity Robust Adaptive Beamforming Based on INCM Reconstruction via Subspace
Projection
Reprinted from: *Sensors* **2021**, *21*, 7783, doi:10.3390/s22145105 . . . . **73**

Zhaonian Wang, Jiangbing Du, Weihong Shen, Jiacheng Liu and Zuyuan He
Efficient Design for Integrated Photonic Waveguides with Agile Dispersion
Reprinted from: *Sensors* **2021**, *21*, 6651, doi:10.3390/s21196651 . . . . **87**

**Mingming Tan, Md Asif Iqbal, Tu T. Nguyen, Paweł Rosa, Lukasz Krzczanowicz, Ian. D.
Phillips, et al.**
Raman Amplification Optimization in Short-Reach High Data Rate Coherent Transmission
Systems
Reprinted from: *Sensors* **2021**, *21*, 6521, doi:10.3390/s21196521 . . . . **101**

**Mingming Tan, Paweł Rosa, Tu T. Nguyen, Mohammad A. Z. Al-Khateeb, Md. Asif Iqbal,
Tianhua Xu, et al.**
Distributed Raman Amplification for Fiber Nonlinearity Compensation in a Mid-Link Optical
Phase Conjugation System
Reprinted from: *Sensors* **2022**, *22*, 758, doi:10.3390/s22030758 . . . . **109**

Tianxu Xu, Dong An, Yuetong Jia and Yang Yue
A Review: Point Cloud-Based 3D Human Joints Estimation
Reprinted from: *Sensors* **2021**, *21*, 1684, doi:10.3390/s21051684 . . . . **127**

MDPI

Editorial

Editorial: Special Issue "Optical Signal Processing Technologies for Communication, Computing, and Sensing Applications"

Jiangbing Du [1,*], Yang Yue [2], Jian Zhao [3] and Yange Liu [4]

1   State Key Laboratory of Advanced Optical Communication Systems and Networks, Shanghai Jiao Tong University, Shanghai 200240, China
2   School of Information and Communications Engineering, Xi'an Jiaotong University, Xi'an 710049, China
3   School of Precision Instruments and Optoelectronics Engineering, Tianjin University, Tianjin 300072, China
4   Institute of Modern Optics, Nankai University, Tianjin 300071, China
*   Correspondence: dujiangbing@sjtu.edu.cn

Citation: Du, J.; Yue, Y.; Zhao, J.; Liu, Y. Editorial: Special Issue "Optical Signal Processing Technologies for Communication, Computing, and Sensing Applications". *Sensors* 2023, 23, 2606. https://doi.org/10.3390/s23052606

Received: 21 February 2023
Accepted: 24 February 2023
Published: 27 February 2023

Optical technology is one of the key technologies that have been widely used for communication, computing and sensing. By utilizing different degrees of freedom for photons, optical signals can be detected and processed in different dimensions, including amplitude, phase, polarization, time, frequency, and spatial mode. Multidimensional signal processing technologies have thus been broadly studied to improve the performance of communication, sensing and even computing systems. Recently, innovative optical signal processing methods and devices have emerged to serve those needs driven by applications including but not limited to optical fiber transmission, supercontinuum generation, phase conjugation, free space optical communication, optical beamforming, photonic integration, fiber amplification, pose estimation and so on.

This Special Issue aims to explore those emerging and enabling technologies of signal processing methods and devices for optical communication, optical computing, and optical sensing. This Special Issue consists of two review papers, one communication, and seven articles.

As for optical signal processing in fiber communications, Y. Zhu et al. presented a proof-of-concept study of stimulated Brillouin scattering (SBS)-induced nonlinear distortion for 10 Gbaud and 28 Gbaud SSB 16QAM transmission over 80 km standard single mode fiber (SSMF) based on a Kramers–Kronig receiver with a significantly reduced bit error rate [1]. M. Tan et al. compared the transmission performances of 600 Gbit/s PM-64QAM WDM signals over 75.6 km of single-mode fiber (SMF) using EDFA, discrete Raman, hybrid Raman/EDFA, and first-order or second-order (dual-order) distributed Raman amplifiers [2]. They also reviewed and studied the designs of distributed Raman amplifiers with respect to nonlinear compensation and bandwidth extension in a mid-link optical phase conjugation system [3,4].

As for optical signal processing for sensing applications, J. Yang et al. designed a silica-cladded Germania-doped ring-core fiber (RCF) that supports orbital angular momentum (OAM) modes. By optimizing the fiber structure parameters, a beyond-two-octave supercontinuum spectrum of the $OAM_{1,1}$ mode can be generated [5]. Regarding 3D human pose estimation, T. Xu et al. reviewed and summarized the recent development on the point cloud-based pose estimation of the human body [6]. The challenges involved and problems to be solved in future studies have also been discussed.

For free space and space division multiplexing applications, C. Álvarez-Roa et al. investigated the application of free space optical (FSO) communications, energy harvesting, and unmanned aerial vehicles (UAVs) as key technology enablers of a cost-efficient backhaul/fronthaul framework for 5G and beyond (5G+) networks [7]. Y. Duan et al. presented a low-complexity robust adaptive beamforming (RAB) method based on an interference-noise covariance matrix (INCM) reconstruction and SOI SV estimation [8].

L. M. Torres et al. studied the linear multiple-input multiple-output (MIMO) receiver designed to optimize the minimum mean square error (MMSE) for a coherent SDM optical communication system, without previous assumptions on receiver oversampling or analog front-end realizations [9].

For the photonic integrated optical signal processing applications, Z. Wang et al. demonstrated a machine learning-based method for agile dispersion engineering of integrated photonic waveguide using a horizontal double-slot structure [10]. Agile dispersion shapes, including broadband low dispersion, constant dispersion and slope-maintained linear dispersion, can be obtained efficiently with high precision.

The Special Issue of optical signal processing only covers few aspects of the powerful and attractive capabilities of optics. Enabling methods, materials, devices, chips and systems for optical signal processing are emerging every day. Optical signal processing powered next-generation communication, computing and sensing can be highly expected.

Conflicts of Interest: The authors declare no conflict of interest.

References

1. Zhu, Y.; Du, J.; Shen, W.; He, Z. Nonlinear Distortion by Stimulated Brillouin Scattering in Kramers-Kronig Receiver Based Optical Transmission. *Sensors* **2022**, *22*, 7287. [CrossRef] [PubMed]
2. Tan, M.; Iqbal, M.A.; Nguyen, T.T.; Rosa, P.; Krzczanowicz, L.; Phillips, I.D.; Harper, P.; Forysiak, W. Raman Amplification Optimization in Short-Reach High Data Rate Coherent Transmission Systems. *Sensors* **2021**, *21*, 6521. [CrossRef] [PubMed]
3. Tan, M.; Rosa, P.; Nguyen, T.T.; Al-Khateeb, M.A.Z.; Iqbal, M.A.; Xu, T.; Wen, F.; Ania-Castañón, J.D.; Ellis, A.D. Distributed Raman Amplification for Fiber Nonlinearity Compensation in a Mid-Link Optical Phase Conjugation System. *Sensors* **2022**, *22*, 758. [CrossRef]
4. Rosa, P.; Martella, G.R.; Tan, M. Bandwidth Extension in a Mid-Link Optical Phase Conjugation. *Sensors* **2022**, *22*, 6385. [CrossRef]
5. Yang, J.; Wang, Y.; Fang, Y.; Geng, W.; Zhao, W.; Bao, C.; Ren, Y.; Wang, Z.; Liu, Y.; Pan, Z.; et al. Over-Two-Octave Supercontinuum Generation of Light-Carrying Orbital Angular Momentum in Germania-Doped Ring-Core Fiber. *Sensors* **2022**, *22*, 6699. [CrossRef] [PubMed]
6. Xu, T.; An, D.; Jia, Y.; Yue, Y. A Review: Point Cloud-Based 3D Human Joints Estimation. *Sensors* **2021**, *21*, 1684. [CrossRef] [PubMed]
7. Álvarez-Roa, C.; Álvarez-Roa, M.; Martín-Vega, F.J.; Castillo-Vázquez, M.; Raddo, T.; Jurado-Navas, A. Performance Analysis of a Vertical FSO Link with Energy Harvesting Strategy. *Sensors* **2022**, *22*, 5684. [CrossRef] [PubMed]
8. Duan, Y.; Yu, X.; Mei, L.; Cao, W. Low-Complexity Robust Adaptive Beamforming Based on INCM Reconstruction via Subspace Projection. *Sensors* **2021**, *21*, 7783. [CrossRef] [PubMed]
9. Torres, L.M.; Cañete, F.J.; Díez, L. Matched Filtering for MIMO Coherent Optical Communications with Mode-Dependent Loss Channels. *Sensors* **2022**, *22*, 798. [CrossRef] [PubMed]
10. Wang, Z.; Du, J.; Shen, W.; Liu, J.; He, Z. Efficient Design for Integrated Photonic Waveguides with Agile Dispersion. *Sensors* **2021**, *21*, 6651. [CrossRef] [PubMed]

Article

Nonlinear Distortion by Stimulated Brillouin Scattering in Kramers-Kronig Receiver Based Optical Transmission

Yuzhu Zhu [1,2], **Jiangbing Du** [2,*], **Weihong Shen** [2] **and Zuyuan He** [2]

[1] AVIC Aeronautical Radio Electronics Research Institute, Shanghai 200241, China
[2] State Key Laboratory of Advanced Optical Communication Systems and Networks, Shanghai Jiao Tong University, Shanghai 200240, China
* Correspondence: dujiangbing@sjtu.edu.cn

Abstract: Nonlinear distortion for single-sideband (SSB) signals will significantly reduce the performance of Kramers–Kronig (KK) receiver-based optical transmission. In this work, we present a proof-of-concept study of stimulated Brillouin scattering (SBS)-induced nonlinear distortion for 10 Gbaud and 28 Gbaud SSB QAM16 transmission over 80 km standard single mode fiber (SSMF) based on a KK receiver. Significantly reduced bit error rate (BER) has been experimentally observed due to the SBS and the threshold of SBS at about 7 dBm is detected for such an 80 km SSMF link. With left sideband (LSB) modulation of SSB, together with optical filtering, reduced SBS nonlinear distortion has been achieved with ~2 dB power tolerance improvement. The results reveal an important issue of SBS-induced nonlinear distortion, which would be of great significance for KK receiver-based optical transmission applications.

Keywords: nonlinear distortion; Kramers–Kronig receiver; stimulated Brillouin scattering

Citation: Zhu, Y.; Du, J.; Shen, W.; He, Z. Nonlinear Distortion by Stimulated Brillouin Scattering in Kramers-Kronig Receiver Based Optical Transmission. *Sensors* **2022**, *22*, 7287. https://doi.org/10.3390/s22197287

Academic Editor: Peter Han Joo Chong

Received: 30 August 2022
Accepted: 24 September 2022
Published: 26 September 2022

Publisher's Note: MDPI stays neutral with regard to jurisdictional claims in published maps and institutional affiliations.

1. Introduction

In recent years, the rapid development of information industries, such as autonomous driving, 5G, and so on, has led to an urgent demand for data transmission, especially for optical interconnection. Intensity modulation/direct detection (IM/DD) is widely used for optical interconnection at conventional 1310 nm and 15,500 nm wavebands, and has recently been extended to the 2 micron waveband [1,2]. Advanced modulations have been proposed for versatile IM/DD systems [3–5], in which nonlinear distortion becomes a very fatal issue due to the much tighter signal-to-noise ratio (SNR) budget, as well as the higher peak-to-average-power-ratio (PAPR). Normally, DD system can only be applied to IM signal for the detection of amplitude information rather than phase information [6]. In 2016, Mecozzi et al. proposed a Kramers–Kronig (KK) self-coherent receiver [7]. On the basis of satisfying the minimum phase condition of signal [8], the phase of the optical field can be restored by its intensity through a KK relationship by using a single photodetector (PD) [9–12]. This scheme has attracted widespread attention since it was proposed. Many studies have been carried out which prove the excellent performance of the KK receiver [13–15].

However, the KK receiver requires that the signal should meet the minimum phase condition, leading to a single sideband modulation (SSB) with a considerably high carrier-to-signal power ratio (CSPR), which could be over 10 dB. Serious nonlinear distortion can be expected for such a SSB signal due to the high CSPR, high PAPR, and tight SNR budget. The high CSPR is needed to maintain a small bit error rate (BER) [16–18]. However, on the other hand, nonlinear distortion by stimulated Brillouin scattering (SBS) will be easily excited for km-level distance optical fiber transmission. Particularly, with a guard band between the carrier and signal, the carrier itself can be considered as a standalone narrow linewidth continuous wave, which makes the SBS threshold even smaller, leading to worse SBS nonlinear distortion. Therefore, the investigation of SBS based nonlinear distortion to KK receiver-based optical fiber transmission is urgently needed.

In this work, we carried out a proof-of-concept study with solid experiments to unfold the nonlinear distortion limitation by SBS for KK signals, which is a unique phenomenon for KK receiver due to the high CSPR property of SSB signal. A significant reduction in the transmission performance due to SBS is observed for 10 Gbaud and 28 Gbaud SSB QAM16 signal transmissions over 80 km SSMF based on a KK receiver. The method of sideband filtering is proposed to reduce nonlinear distortion by SBS. Those results are of increasing value to researchers in this field, along with the increased use of a KK receiver in optical communication applications, which offer a higher data rate and long transmission distance.

2. Nonlinear Distortion Due to SBS for SSB Signals

Figure 1 shows the schematic mechanism of SBS induced nonlinear distortion for SSB signals. The generation of a SSB signal can be realized by standard in-phase and quadrature (IQ) modulation, optical filtering, or carrier-signal offset combination, which is adopted in this work. As depicted by Figure 1a,c, left single-sideband (LSB) and right single-sideband (RSB) SSB signals can be flexibly obtained. The SSB signal can be realized with or without guard band between the carrier and the signal. Improved optical detection can be obtained for SSB with guard band, as shown by Figure 1e,g.

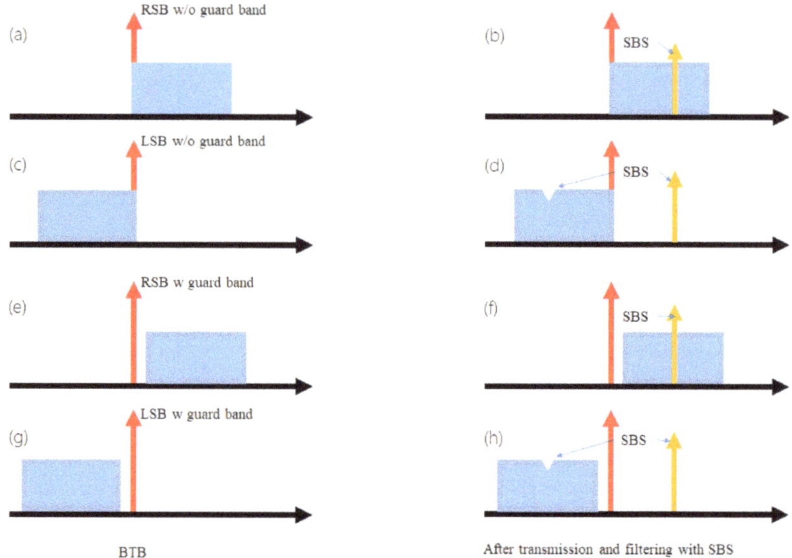

Figure 1. SBS induced nonlinear distortion for SSB signals. (**a**) RSB without guard band; (**b**) RSB without guard band after SBS distortion; (**c**) LSB without guard band; (**d**) LSB without guard band after SBS distortion; (**e**) RSB with guard band; (**f**) RSB with guard band after SBS distortion; (**g**) LSB with guard band; (**h**) LSB with guard band after SBS distortion.

As an important nonlinear distortion, SBS can be easily induced due to its comparably low threshold. In conventional applications, the signals are modulated with a high data rate without significant carrier power left and, thus, lead to neglected SBS. However, for the KK receiver, the SSB signal is needed with a high carrier power in order to maintain a small detection error. Typical CSPR exceeds 10 dB, which makes the SBS issue no longer neglectable, as shown in Figure 1b,d. In particular, as for SSB signals with guard band, the carrier is in fact a standalone laser with narrow linewidth. The SBS threshold would be even lower, and significant nonlinear distortion due to SBS can be expected, as schematically shown in Figure 1f,h.

3. Experimental Setup

The experimental setup for the KK receiver-based optical transmission is built for the investigation of the SBS induced nonlinear distortion for SSB signals, as shown in Figure 2.

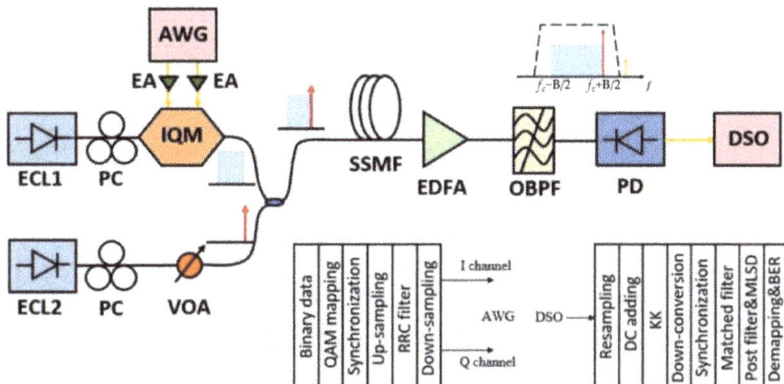

Figure 2. Experiment setup and DSP. Here, ECL: external cavity laser; PC: polarization controller; IQM: in-phase and quadrature modulator; AWG: arbitrary waveform generator; VOA: variable optical attenuator; EDFA: erbium-doped fiber amplifier; OBPF: optical bandpass filter; DSO: digital storage oscilloscope.

At the transmitter, a pseudo-random bit stream ($2^{17}-1$) is generated by a 54 GSa/s arbitrary waveform modulator (AWG), which is mapped into a QAM16 signal after encoding. The baud rate R_s is set to 10 Gbaud and 28 Gbaud, respectively. Then, the signal is pulse-shaped with a root-raised cosine (RRC), and the roll-off coefficient $\alpha = 0.01$. The signal spectrum bandwidth is $B_s = (1 + \alpha)R_s$. A continuous wave light from an external cavity laser (ECL1) at 1550 nm is fed into the IQ modulator. Additionally, the other ECL (ECL2, with a linewidth of 15 KHz) generates a continuous wave (CW) as an optical carrier, whose power can be adjusted to obtain a different output total power and CSPR. The optical SSB QAM16 signal is then launched into the 80 km standard single mode-fiber (SSMF) at 1550 nm.

At the receiver, the transmitted optical field is first detected by the PD with a 3 dB bandwidth of ~22 GHz, and the electrical signal is captured by an 80 GSa/s digital storage oscilloscope (DSO, Keysight DSOZ592A). As shown by the inset of Figure 2, an optical bandpass filter (OBPF) before PD can be used for the filtering of the SBS influence, so as to reduce the nonlinear distortion.

Since the nonlinear square root operation and logarithmic operation of the KK algorithm will broaden the frequency spectrum of the signal, the signal is up-sampled before the KK algorithm processing, and the sampling rate is generally set to 4 [7]. Then, the KK receiver algorithm is used to recover the amplitude and phase of the signal, and the detail is shown in Figure 3. Due to the frequency interval between the carrier and the signal, the signal is an intermediate frequency signal at this time, and it needs to be down-converted to demodulate the baseband signal. To mitigate the inter-symbol interference caused by the limited bandwidth, a feedforward equalizer (FFE) is applied. However, the FFE boosts the in-band noise, which reduces the in-band signal-to-noise ratio (SNR). Therefore, a maximum likelihood sequence decision (MLSD) equalizer consisting of a post-filter and a conversional MLSD is employed. The post-filter is a two-tap finite impulse response filter. In addition, post-dispersion compensation is used in DSP for the 80 km SSMF transmission. After recovering the optical field with a KK receiver, the time domain information of the signal is transformed to the frequency domain by the fast Fourier transform (FFT) technique, and then the phase shift due to dispersion is converted back. Finally, the signal is converted to the time domain by the inverse fast Fourier transform (IFFT) technique.

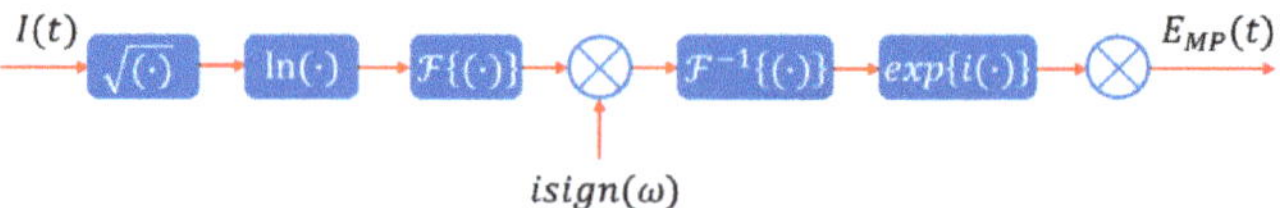

Figure 3. Signal processing of the KK receiver scheme.

4. Results and Discussions

At a launch power of 11 dBm, corresponding to a CSPR of 13 dB, SBS can be easily observed. Figure 4 shows the optical spectra of the 10 Gbaud and 28 Gbaud QAM16 SSB signals after the 80 km SSMF transmission. From Figure 4a,d, one can clearly find the Stokes component induced by SBS at 1550.24 nm and 1550.27 nm for 10 Gbaud and 28 Gbaud QAM16, respectively, which exactly corresponds to an SBS frequency shift of 10.8 GHz with respect to the carrier wavelength. The anti-Stokes component will lead to attenuation at the shorter wavelength side with the same frequency shift. At 10 Gbaud, the anti-Stokes component is outside of the signal. The anti-Stokes component is completely buried in the LSB signal at 28 Gbaud. Therefore, it is unable to be observed from the spectra.

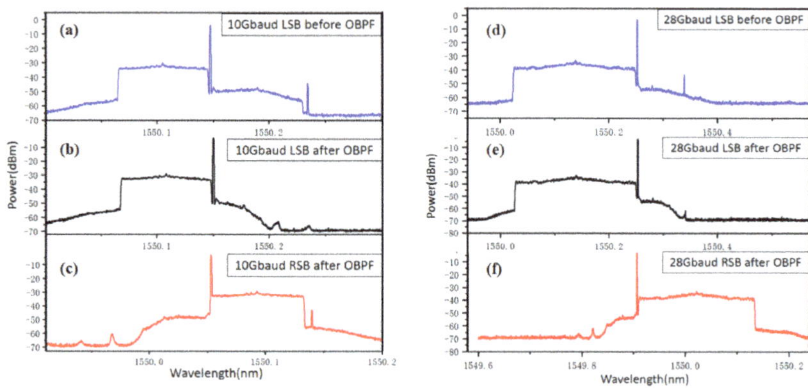

Figure 4. Optical spectra of the 10 Gbaud and 28 Gbaud QAM16 signals. (**a**) 10 Gbaud LSB signal before OBPF; (**b**) 10 Gbaud LSB signal after OBPF; (**c**) 10 Gbaud RSB signal after OBPF; (**d**) 28 Gbaud LSB signal before OBPF; (**e**) 28 Gbaud LSB signal after OBPF; (**f**) 28 Gbaud RSB signal after OBPF.

Using a flattop OBPF with high roll-off, one can mitigate the SBS induced nonlinear distortion by removing the Stokes component from the signal, as shown in Figure 4b,e for LSB signals. As for the RSB signal at a lower speed, such as 10 Gbaud in Figure 4c, the Stokes component would be too close to the signal and, thus, cannot be filtered out. As for the RSB signal at higher speeds, such as 28 Gbaud in Figure 4f, the Stokes component is also completely buried in the signal and, thus, cannot be filtered out either. Therefore, it can be expected that LSB suffers less SBS nonlinear distortion compared with RSB after the OBPF filtering.

Figure 5 shows the BER of the 10 Gbaud signal for back-to-back (B2B) and after 80 km SSMF transmission. One can observe from Figure 5a that the BER performances are at the same level and vary within a very small range along with the increase in the launch power for B2B. This is reasonable, since there is not yet any nonlinear distortion, and the CSPR is already large enough for the KK receiver to obtain an optimal BER. However, in Figure 5b, after 80 km SSMF transmission, significantly increased BER can be observed when the launch power exceeds ~7 dBm, which is mainly due to the existence of SBS. This is due to several reasons. The first is that the increased launch power is purely from the carrier from ECL2, with signal power remaining unchanged for the branch of ECL1 in Figure 2. Thus, CSPR increases accordingly, which should lead to reduced BER without nonlinear distortion. Secondly, signal power remains unchanged, which means that the nonlinear distortion due to Kerr effect by the signal is also maintained. Thirdly, the increase in BER is mainly induced by the increased carrier power, which can only be explained by the SBS induced nonlinear distortion.

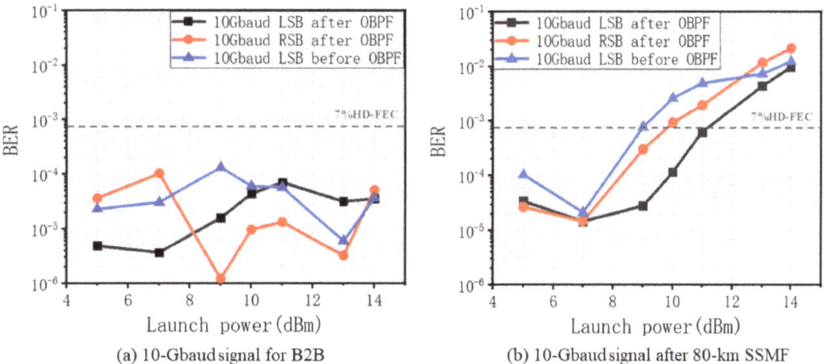

(a) 10-Gbaud signal for B2B   (b) 10-Gbaud signal after 80-km SSMF

Figure 5. BER versus launch power. (**a**) 10 Gbaud for B2B; (**b**) after 80 km SSMF transmission.

As previously discussed, the nonlinear distortion caused by SBS can be mitigated with the use of OBPF. The conclusion can also be proved by the BER results in Figure 5b, which presents reduced BER for the signals after OBPF. Particularly, one can also observe reduced BER for LSB with respect to RSB, which is in good accordance with the principle. There is a ~2 dB power tolerance improvement for LSB with OBPF filtering with respect to the error-free BER of 3.8×10^3.

The nonlinear distortion can also be observed from the constellations, as depicted in Figure 6, at different launch powers of 5 dBm, 11 dBm, and 14 dBm. For both RSB and LSB, significantly increased noise rather than constellation shape distortion can be found in Figure 6b,c,e,f. This is also a proof of SBS rather than Kerr for the nonlinear distortion. Meanwhile, one can also observe slightly improved constellation noise performance for LSB at 14 dBm compared with that of RSB, which is in good agreement with previous predictions as well as the BER results.

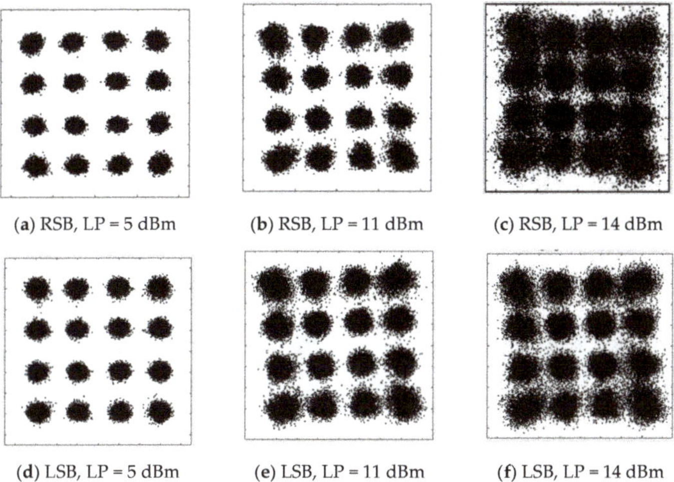

(a) RSB, LP = 5 dBm   (b) RSB, LP = 11 dBm   (c) RSB, LP = 14 dBm

(d) LSB, LP = 5 dBm   (e) LSB, LP = 11 dBm   (f) LSB, LP = 14 dBm

Figure 6. Constellation of 10 Gbaud QAM16 after 80 km SSMF transmission. (**a**) RSB, LP = 5 dBm; (**b**) RSB, LP = 11 dBm; (**c**) RSB, LP = 14 dBm; (**d**) LSB, LP = 5 dBm; (**e**) LSB, LP = 11 dBm; (**f**) LSB, LP = 14 dBm.

Similar performance can be observed for SSB signals at the 28 Gbaud data rate. The BER curves are shown in Figure 7. As for the B2B case in Figure 7a, BERs are maintained at almost the same level, since the CSPR is large enough with the absence of nonlinear distortion. The increased BER can be found in Figure 7b when the launch power exceeds 7 dBm, which is the same as that of the 10 Gbaud signals. This is also reasonable, since the threshold of the SBS should be the same for different data rates, which is one more proof of the SBS nonlinear distortion. The mitigation of the SBS nonlinear distortion can also be achieved by using the OBPF filtering with reduced BER for both LSB and RSB. However, there is no significant difference between LSB and RSB. We believe that this is due to the bandwidth limitation of the signal channel for 28 Gbaud, which to a certain extent erases their performance difference. This can also be observed from the performance of the constellations, as shown in Figure 8, in which the noise difference between LSB and RSB at 14 dBm is much smaller for 28 Gbaud than that for 10 Gbaud.

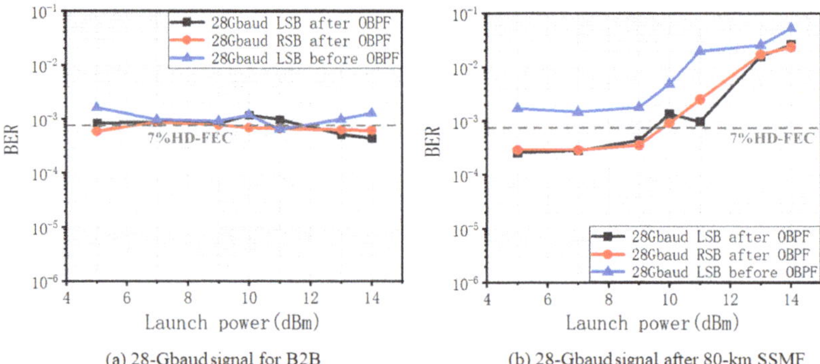

Figure 7. BER versus launch power. (**a**) 28 Gbaud for B2B; (**b**) after 80 km SSMF transmission.

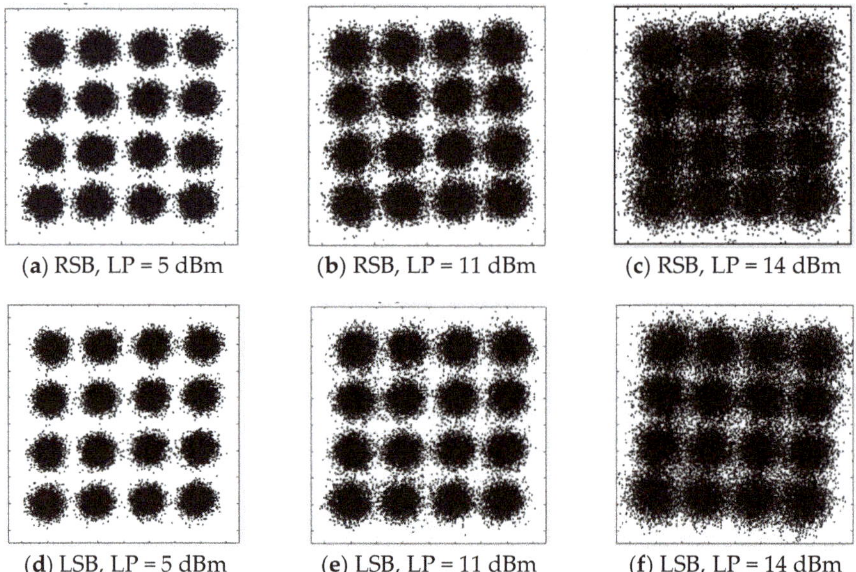

Figure 8. Constellation of 28 Gbaud QAM16 at 80 km SSMF transmission. (**a**) RSB, LP = 5 dBm; (**b**) RSB, LP = 11 dBm; (**c**) RSB, LP = 14 dBm; (**d**) LSB, LP = 5 dBm; (**e**) LSB, LP = 11 dBm; (**f**) LSB, LP = 14 dBm.

On the whole, SBS has different impacts on signals with varying data rates, and some of the impacts of the nonlinear distortion by SBS as CSPR increased can be alleviated by utilizing OBPF to filter out the component of SBS. The SBS has a greater overall impact on the higher data rate transmission systems than it does on the lower data rate systems when OBPF can be applied to filter out peak of SBS. For the higher rate transmission system, the impact of SBS on the RSB signal is greater than that on the LSB signal under the same experimental conditions.

5. Conclusions

In this work, the SBS induced nonlinear distortion for 10 Gbaud and 28 Gbaud SSB QAM16 transmission over an 80 km SSMF based on a KK receiver is experimentally observed and investigated for the first time. Significantly reduced BER can be induced along with the increase in CSPR due to SBS, which makes SBS an unneglectable issue for KK receiver-based systems. An SBS threshold of about 7 dBm is detected for such an 80 km SSMF link. Different nonlinear distortion performances are found for LSB and RSB signals, and LSB performs better than RSB. Here, a ~2 dB power tolerance improvement is achieved by LSB with optical filtering to remove the SBS. The proof-of-concept results in this work unfold an important issue of SBS-induced nonlinear distortion, which would be of great significance for KK receiver-based optical transmission applications.

Author Contributions: Conceptualization, J.D. and Y.Z.; methodology, J.D. and Y.Z.; experiment investigation, Y.Z. and W.S.; writing—original draft preparation, Y.Z.; writing—review and editing, J.D. and Z.H.; supervision, J.D. and Z.H. All authors have read and agreed to the published version of the manuscript.

Funding: This work is supported by National Key R&D Program of China (2018YFB1801804), National Natural Science Foundation of China (NSFC) (61935011, 61875124).

Institutional Review Board Statement: Not applicable.

Informed Consent Statement: Not applicable.

Data Availability Statement: Not applicable.

Acknowledgments: Thanks to all the reviewers and editors for their careful work and pertinent suggestions.

Conflicts of Interest: The authors declare no conflict of interest.

References

1. Shen, W.; Du, J.; Sun, L.; Wang, C.; Xu, K.; Chen, B.; He, Z. 100-Gbps 100-m Hollow-Core Fiber Optical Interconnection at 2-Micron Waveband by PS-DMT. In Proceedings of the 2020 Optical Fiber Communications Conference and Exhibition (OFC), San Diego, CA, USA, 8–12 March 2020; pp. 1–3.
2. Shen, W.; Du, J.; Wang, C.; Sun, L.; Xu, K.; He, Z. Single lane 90-Gbps optical interconnection at 2-micron waveband. In Proceedings of the 2019 24th OptoElectronics and Communications Conference (OECC) and 2019 International Conference on Photonics in Switching and Computing (PSC), Fukuoka, Japan, 7–11 July 2019; pp. 1–3. [CrossRef]
3. He, Z.; Du, J.; Shen, W.; Huang, Y.; Wang, C.; Xu, K.; He, Z. Inverse Design of Few-Mode Fiber by Neural Network for Weak-Coupling Optimization. In Proceedings of the 2020 Optical Fiber Communications Conference and Exhibition (OFC), San Diego, CA, USA, 8–12 March 2020; pp. 1–3.
4. Shen, W.; Du, J.; Shen, W.; Sun, L.; Wang, C.; Zhu, Y.; Xu, K.; Chen, B.; He, Z. Low-Latency and High-Speed Hollow-Core Fiber Optical Interconnection at 2-Micron Waveband. *J. Light. Technol.* **2020**, *38*, 3874–3882. [CrossRef]
5. Wang, C.; Du, J.; Shen, W.; Chen, G.; Wang, H.; Sun, L.; Xu, K.; Liu, B.; He, Z. QAM classification methods by SVM machine learning for improved optical interconnection. *Opt. Commun.* **2019**, *444*, 1–8. [CrossRef]
6. Kao, K.C.; Hockham, G.A. Dielectric-fibre surface waveguides for optical frequencies. *IEE Proc. J Optoelectron.* **1986**, *133*, 191–198. [CrossRef]
7. Mecozzi, A.; Antonelli, C.; Shtaif, M. Kramers–Kronig coherent receiver. *Optica* **2016**, *3*, 1220–1227. [CrossRef]
8. Kronig, R.L. On the theory of the dispersion of x-rays. *J. Opt. Soc. Am.* **1926**, *12*, 547–557. [CrossRef]

9. Sun, C.; Che, D.; Shieh, W. Comparison of Chromatic Dispersion Sensitivity between Kramers-Kronig and SSBI Iterative Cancellation Receiver. In Proceedings of the 2018 Optical Fiber Communications Conference and Exposition (OFC), San Diego, CA, USA, 11–15 March 2018; pp. 1–3.
10. Vier, E.; Lemmen, R.; Schwarz, H. Energy- and phase-frequency characteristics of nonlinear systems. In Proceedings of the 1997 European Control Conference (ECC), Brussels, Belgium, 1–7 July 1997; pp. 1146–1151. [CrossRef]
11. Civalleri, P. On the formal theory of nonuniform transmission lines. *IEEE Trans. Circuit Theory* **1971**, *18*, 479–481. [CrossRef]
12. Voelcker, H. Demodulation of Single-Sideband Signals Via Envelope Detection. *IEEE Trans. Commun.* **1966**, *14*, 22–30. [CrossRef]
13. Chen, X.; Antonelli, C.; Chandrasekhar, S.; Raybon, G.; Sinsky, J.; Mecozzi, A.; Shtaif, M.; Winzer, P. 218-Gb/s single-wavelength, single-polarization, single-photodiode transmission over 125-km of standard singlemode fiber using Kramers-Kronig detection. In Proceedings of the 2017 Optical Fiber Communications Conference and Exhibition (OFC), Los Angeles, CA, USA, 19–23 March 2017; pp. 1–3.
14. Le, S.T.; Schuh, K.; Chagnon, M.; Buchali, F.; Dischler, R.; Aref, V.; Engenhardt, K.M. 1.72-Tb/s Virtual-Carrier-Assisted Direct-Detection Transmission Over 200 km. *J. Light. Technol.* **2017**, *36*, 1347–1353. [CrossRef]
15. Chen, X.; Cho, J.; Chandrasekhar, S.; Winzer, P.; Antonelli, C.; Mecozzi, A.; Shtaif, M. Single-wavelength, single-polarization, single- photodiode kramers-kronig detection of 440-Gb/s entropy-loaded discrete multitone modulation transmitted over 100-km SSMF. In Proceedings of the 2017 IEEE Photonics Conference (IPC) Part II, Orlando, FL, USA, 1–5 October 2017; pp. 1–2. [CrossRef]
16. An, S.; Zhu, Q.; Li, J.; Su, Y. Accurate Field Reconstruction at Low CSPR Condition Based on a Modified KK Receiver With Direct Detection. *J. Light. Technol.* **2019**, *38*, 485–491. [CrossRef]
17. Zhang, Q.; Shu, C. Frequency Comb Based Kramers-Kronig Detection. In Proceedings of the 2018 IEEE Photonics Conference (IPC), Reston, VA, USA, 30 September–4 October 2018; pp. 1–2. [CrossRef]
18. Zhang, Q.L.; Zheng, B.F.; Shu, C. Kramers–Kronig detection with Brillouin-amplified virtual carrier. *Opt. Lett.* **2018**, *43*, 1367–1370. [CrossRef] [PubMed]

Article

Over-Two-Octave Supercontinuum Generation of Light-Carrying Orbital Angular Momentum in Germania-Doped Ring-Core Fiber

Jian Yang [1], Yingning Wang [1], Yuxi Fang [1], Wenpu Geng [1], Wenqian Zhao [1], Changjing Bao [2], Yongxiong Ren [2], Zhi Wang [1], Yange Liu [1], Zhongqi Pan [3] and Yang Yue [4],*

1   Institute of Modern Optics, Nankai University, Tianjin 300350, China
2   Department of Electrical Engineering, University of Southern California, Los Angeles, CA 90089, USA
3   Department of Electrical & Computer Engineering, University of Louisiana at Lafayette, Lafayette, LA 70504, USA
4   School of Information and Communications Engineering, Xi'an Jiaotong University, Xi'an 710049, China
*   Correspondence: yueyang@xjtu.edu.cn

Abstract: In this paper, we design a silica-cladded Germania-doped ring-core fiber (RCF) that supports orbital angular momentum (OAM) modes. By optimizing the fiber structure parameters, the RCF possesses a near-zero flat dispersion with a total variation of $<\pm30$ ps/nm/km over 1770 nm bandwidth from 1040 to 2810 nm for the $OAM_{1,1}$ mode. A beyond-two-octave supercontinuum spectrum of the $OAM_{1,1}$ mode is generated numerically by launching a 40 fs 120 kW pulse train centered at 1400 nm into a 12 cm long designed 50 mol% Ge-doped fiber, which covers 2130 nm bandwidth from 630 nm to 2760 nm at -40 dB of power level. This design can serve as an efficient way to extend the spectral coverage of beams carrying OAM modes for various applications.

Keywords: orbital angular momentum; nonlinear optics; supercontinuum

Citation: Yang, J.; Wang, Y.; Fang, Y.; Geng, W.; Zhao, W.; Bao, C.; Ren, Y.; Wang, Z.; Liu, Y.; Pan, Z.; et al. Over-Two-Octave Supercontinuum Generation of Light-Carrying Orbital Angular Momentum in Germania-Doped Ring-Core Fiber. *Sensors* **2022**, *22*, 6699. https://doi.org/10.3390/s22176699

Academic Editor: Nicolas Riesen

Received: 7 June 2022
Accepted: 8 July 2022
Published: 5 September 2022

Publisher's Note: MDPI stays neutral with regard to jurisdictional claims in published maps and institutional affiliations.

1. Introduction

Orbital angular momentum (OAM) has gained widespread attention due to its twisted helical phase front and doughnut-shaped intensity distribution. OAM beams have theoretically infinite topological states and unique phase singularity. Therefore, OAM has been applied to a variety of cutting-edge technologies, such as optical communication systems [1–6], super-resolution microscopy [7,8], optical sensing [9–11], laser material processing [12,13] and imaging [14–17]. Especially in the field of communication, OAM beams with different topological numbers can form an orthogonally modal set, which can be effectively multiplexed and demultiplexed. Therefore, we can transmit multiple independent beams simultaneously in the same space and frequency band. The spectral efficiency and data capacity of the communication systems can thus be improved by using OAM mode division multiplexing (MDM), since coaxial beams with different OAM states can be efficiently separated. Meanwhile, many methods for the generation of OAM beams have been proposed and experimentally demonstrated by using different types of converters [18,19]. It is of great significance to effectively maintain the OAM beams propagating in the optical fiber. Unfortunately, the OAM beams cannot propagate stably in a conventional fiber, in which the quasi-degenerate modes can be easily coupled to each other. Ring-core fiber (RCF) could potentially solve the problem, as the OAM beams have a similarly ring-shaped intensity profile [20]. OAM beams can be generated from Gaussian beams, but a single OAM beam still has a narrow wavelength range, which limits its application in many fields. Therefore, supercontinuum (SC) generation of OAM beams spanning thousands of nanometers wavelength range is of great significance for applications that require broadband OAM beams.

SC has important applications in many fields, such as optical communication [21], optical frequency comb [22,23] and optical coherence tomography [24,25], which has made it an active research field for decades. One of the most important applications of SC in optical communications is to serve as a multi-wavelength source for ultra-broadband WDM systems [26]. In addition, SC is widely used for all-optical analog-to-digital conversion [27] and TDM-to-WDM-to-TDM conversion [28]. Launching ultrashort pulse with an optical vortex state has been demonstrated to generate SC maintaining optical vortex properties [29]. Moreover, one pioneer research has indicated that an octave-spanning SC of OAM beams could be generated through specially designed optical fibers [30]. In particular, the spectral range of the SC is the key factor to be considered. Recently, SC carrying OAM in fiber has been reported to effectively expand the spectral coverage of the OAM beams [31–34]. However, most of the core materials used were As_2S_3, which is a toxic and harmful substance. In addition, it is not trivial to process and manufacture optical fibers composed of this material [35]. Thus, we choose to use Germania as the core material, which has been widely used in optical fiber manufacturing.

For decades, Germania-doped silica has been one of the most common materials for optical fiber communication due to its excellent physical properties and compatibility with silica glass. Germania-doped silica features lots of excellent properties, such as high mechanical strength, long-term structural stability, low sensitivity to ionizing radiation, low chemical activity, close thermal expansion, etc. These characteristics make it possible to fabricate optical fibers with good geometrical quality and long operation life. In addition, the fabrication of Germania-doped silica fibers has also been demonstrated to be feasible in previous experimental studies [36,37]. Furthermore, this material possesses both low intrinsic absorption and Rayleigh scattering in the near-infrared (IR) spectral range, which leads to very low loss over a wide spectral range.

In this work, we design a 50 mol% Ge-doped ring-core silica fiber supporting OAM modes. We tune its structural parameters to 1 μm inner SiO_2 radius and 1.8 μm Ge-doped ring width, making the ring-core fiber possessing flat and low dispersion with a small variation of $<\pm30$ ps/nm/km from 1040 to 2810 nm. A 70 fs 120 kW secant hyperbolic pulse train with the central wavelength of 1400 nm is chosen to be the input source, which is finally broadened into an over-two-octave SC spectrum carrying the $OAM_{1,1}$ mode with the wavelength from 630 nm to 2760 nm at -40 dB of power level by propagating through a 12 cm long designed 50 mol% Ge-doped ring-core silica fiber. COMSOL Multiphysics software is used to simulate the OAM mode properties, and MATLAB-based generalized pulse-propagation equation is used to simulate the process of SC generation.

2. Concept and Fiber Structure

The schematic diagram of SC generation for the OAM mode is illustrated in Figure 1a. When an intense ultrashort pulse is launched into the optical fiber, the spectrum is widened because of the interaction between the dispersion of the transmission medium and various nonlinear effects. The cross-section of the fiber with a low-index inner SiO_2 substrate, a high-index Ge-doped ring and a SiO_2 cladding is shown in Figure 1b.

The material index difference between the silica cladding and Germania-doped annular region is large enough, leading to a large effective refractive index separation between the adjacent modes, which guarantees the modal separation and reduces the corresponding modal crosstalk. Furthermore, this design substantially reduces the modal coupling. Therefore, the choice of fiber materials and the design of fiber structure can perfectly support the propagation of OAM modes. Moreover, the Germania-doped materials have higher nonlinear coefficients than silica over a wide transparent window close to the near-infrared, potentially enabling efficient SC generation in a broad spectral range for the OAM mode. Additionally, we set the cladding diameter to 125 μm, which is identical to the standard single-mode fiber (SMF). Here, our investigation focuses on the SC generation of $HE_{2,1}$, corresponding to the $l = 1$ OAM mode, which is composed of $HE_{2,1}^{even} + i \times HE_{2,1}^{odd}$.

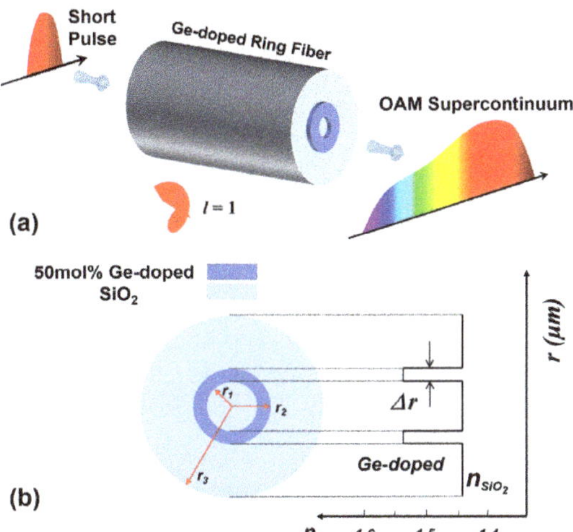

Figure 1. (**a**) Schematic diagram of SC generation; (**b**) Cross-section of the Ge-doped silica fiber.

With respect to fabrication, the selection of material and the design of the RCF structure are realistic, and RCFs composed of SiO$_2$ and Ge-doped SiO$_2$ have been manufactured in practice [38], which has shown good performance on the transmission of OAM modes, especially for the OAM$_{1,1}$ mode.

3. Fiber Properties and Dispersion Optimization

In optical waveguides, SC generation is closely related to dispersion conditions. A flat and low dispersion is preferable to achieve SC generation over a wide bandwidth. We calculate the dispersion D by Equation (1), which is in the unit of ps/nm/km [39].

$$D = -\frac{\lambda}{c}\frac{d^2 n_{eff}}{d\lambda^2},\tag{1}$$

where c is the velocity of light in free space, and n_{eff} is the effective refractive index of the OAM mode propagating in the designed RCF. To optimize the designed Ge-doped RCF with flat and low dispersion, we investigate the structure parameters, including the doping concentration of Germania, ring width (Δr) and SiO$_2$ ring radius (r_1).

The doping concentration of Ge is first optimized, and the dispersion curve of OAM$_{1,1}$ mode is shown in Figure 2. One can see that the dispersion curve moves up with the increase in doping concentration under conditions of the same fiber geometric structure, while the smaller doping concentration leads to a faster-changing dispersion. All the dispersion curves for fibers with different doping concentrations have similar trends, which increase first and then decrease over the wavelength.

It is worth noting that higher doping concentration enables the RCF to support the OAM$_{1,1}$ mode over a wider wavelength range. As higher doping concentration makes the refractive index difference between the ring core and the cladding higher, the OAM mode will have a larger cut-off wavelength with a higher doping concentration, which will directly affect the upper wavelength limit of the SC generation. Finally, the Ge-doped concentration is set to 50 mol%, which can potentially reach the target of generating SC with more than two-octave spectral broadening.

Figure 3a illustrates the effect of different ring widths (Δr) on the HE$_{2,1}$ mode dispersion curve. Under the conditions of the same inner silica radius, we can clearly see that the dispersion curve rises as the ring width increases, and the smaller ring width leads to a faster-changing dispersion. Furthermore, we optimize the inner SiO$_2$ radius (r_1), which has less effect on the dispersion, as shown in Figure 3b. The upward trend of the dispersion curve becomes more obvious as r_1 increases.

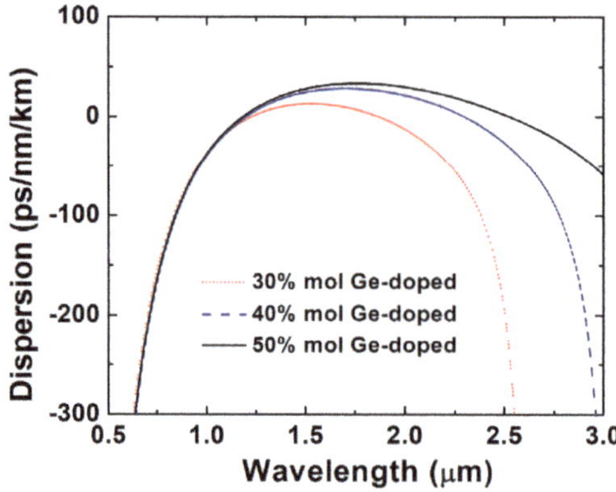

Figure 2. Dispersion-wavelength curves at different Ge-doping concentrations.

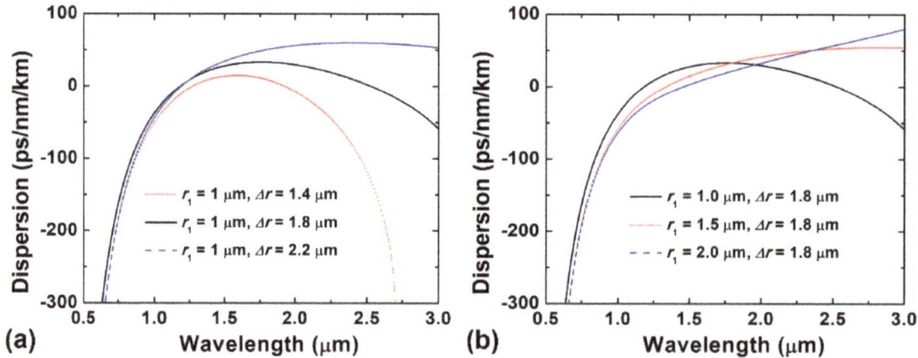

Figure 3. Dispersion-wavelength curves for different (**a**) ring width (Δr) and (**b**) fiber inner SiO$_2$ radius (r_1).

The fiber is relatively dispersive in the short wavelength range due to large material index change. The solid black line in the figure represents the dispersion from 500 nm to 3000 nm of the HE$_{2,1}$ mode in the designed RCF with optimized structure parameters (r_1 =1 µm, Δr =1.8 µm). It can be clearly seen that the optimized RCF structure has a flat and near-zero dispersion profile over a wide wavelength range in the near-infrared region with a total dispersion variation of <±30 ps/nm/km over a 1770 nm bandwidth from 1040 nm to 2810 nm.

Figure 4 depicts the intensity and phase distributions of the OAM$_{1,1}$ mode under different wavelengths supported in the designed RCF. The upper and the lower color bars represent the normalized mode field intensity and the phase change of the OAM$_{1,1}$ mode, respectively. These results are calculated through full-vector finite-element method (FEM). One can notice in Figure 4 that the intensity distributions of the OAM$_{1,1}$ mode maintain annular shape constantly, which is well confined within the Ge-doped ring of the fiber, and the effective mode field area increases with wavelength. Meanwhile, the OAM$_{l,1}$ mode shows a $2l\pi$ phase change azimuthally. The azimuthal phase variation of OAM$_{1,1}$ is 2π, corresponding to a topological charge number of 1.

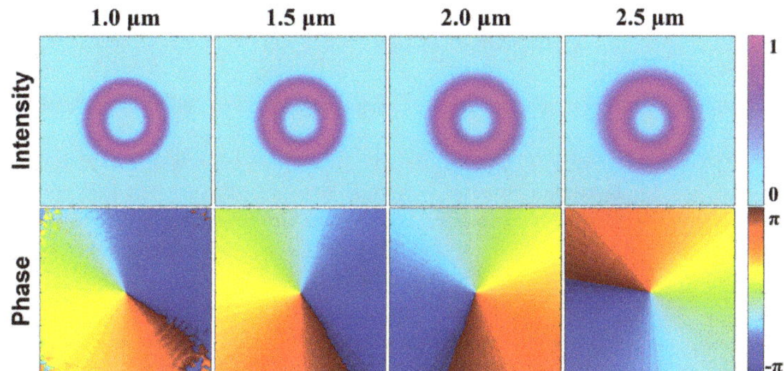

Figure 4. Normalized intensity and phase distributions at different wavelengths of $OAM_{1,1}$ mode supported in the designed RCF.

In the SC generation process, loss is one of the most significant factors. The fiber loss shown in Figure 5a is determined by calculating the imaginary part of the effective refractive index for the $OAM_{1,1}$ mode [40] according to the material loss of silica and Ge-doped silica [41,42]. As the $OAM_{1,1}$ mode cannot be supported normally in the designed RCF for a wavelength larger than 3400 nm, the fiber loss is set to infinity for wavelengths beyond 3400 nm. Nonlinearity is another important parameter that affects the efficiency of the nonlinear process. The nonlinear coefficient (γ) and effective mode area (A_{eff}) of the $OAM_{1,1}$ mode are displayed in Figure 5b. The nonlinear coefficient decreases as the effective mode area increases with the wavelength, as illustrated in Figure 4.

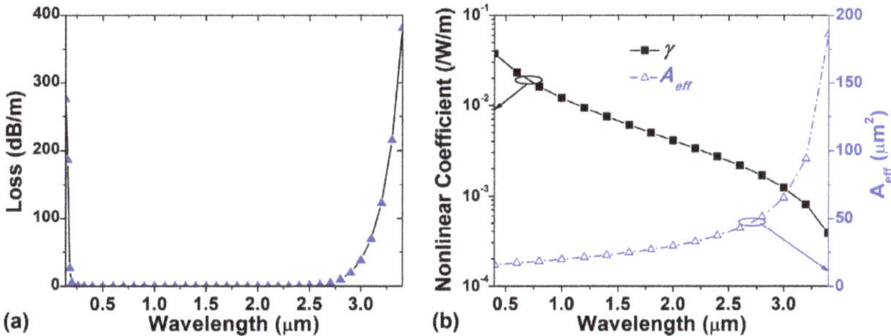

Figure 5. (a) Loss, (b) nonlinear coefficient and effective mode field area (A_{eff}) of the RCF with the optimized design.

4. Supercontinuum Generation

SC generation is simulated numerically using the generalized pulse-propagation equation, which takes into account the contributions of both the linear effects (dispersion and loss) and the nonlinear effects (Kerr nonlinearity effect, self-steepening effect, etc.) [39].

$$\frac{\partial A}{\partial z} + \frac{1}{2}\left(\alpha(\omega_0) + i\alpha_1\frac{\partial}{\partial t}\right)A - i\sum_{n=1}^{\infty}\frac{i^n\beta_n}{n!}\frac{\partial^n A}{\partial t^n} = i\gamma\left(1 + \frac{i}{\omega_0}\frac{\partial}{\partial t}\right)\left(A(z,t)\int_0^{\infty}R(\tau)|A(z,t-\tau)|^2\partial\tau\right) \quad (2)$$

where A is the electric field envelope, α is the loss coefficient, ω_0 is the input pulse frequency, β_n is the group velocity dispersion (GVD), and n is up to 10 in our simulation, τ is the present time frame, $R(t)$ is the nonlinear response function. The functional form of $R(t)$ can be expressed as [39]

$$R(t) = (1 - f_R)\delta(t) + f_R\left(\tau_1^{-2} + \tau_2^{-2}\right)\tau_1 \exp(-t/\tau_2)\sin(t/\tau_1) \tag{3}$$

where f_R represents the fractional contribution of the delayed Raman response. The Raman response function coefficients used in Equation (2) are $f_R = 0.18$, $\tau_1 = 12.2$ fs, $\tau_2 = 83$ fs, respectively [43]. The nonlinear refractive index n_2 for 50 mol% Ge-doped silica is 3.81×10^{-20} m^2/W [44], and a full-vector model is used to obtain the Kerr nonlinear coefficient in the simulation [45,46].

The generated supercontinua in the designed fiber with 1 μm inner SiO$_2$ radius and 1.8 μm ring width are shown in Figure 6. We further analyze several key influence factors, including the pump center wavelength, pump peak power and pulse width. The material refractive indices of silica and 50 mol% Ge-doped silica are obtained using the Sellmeier equations in our model [47,48].

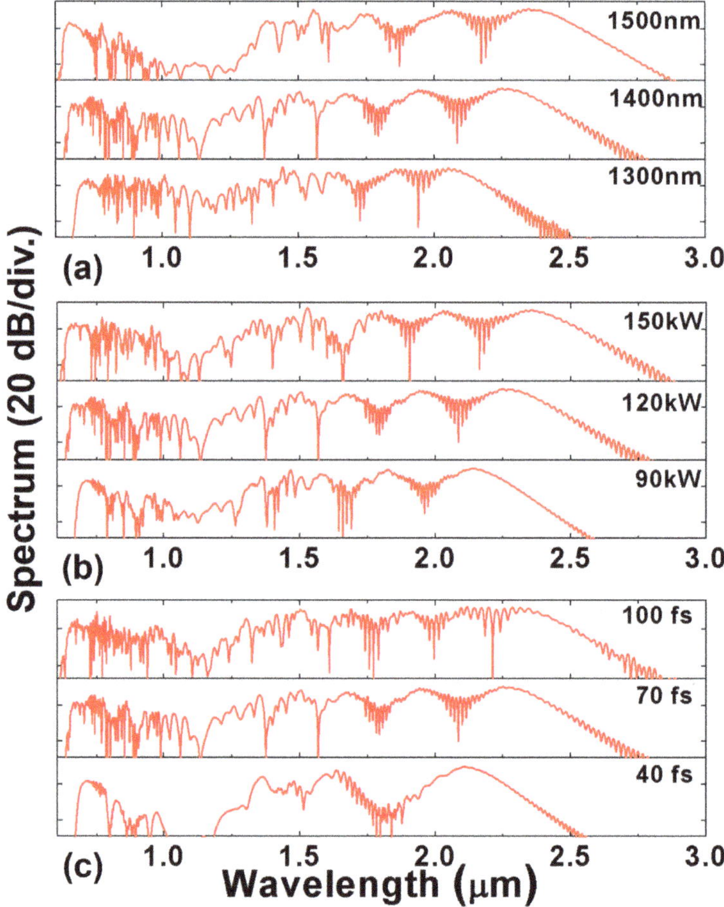

Figure 6. Influence of the initial pulse parameters on SC generation of OAM$_{1,1}$ after a 12 cm designed Ge-doped RCF for different (**a**) λ_0 ($P_0 = 120$ kW, $T_{FWHM} = 70$ fs); (**b**) P_0 ($\lambda_0 = 1400$ nm, $T_{FWHM} = 70$ fs); (**c**) T_{FWHM} ($\lambda_0 = 1400$ nm, $P_0 = 120$ kW).

The influence of the input center wavelength (λ_0) is illustrated in Figure 6a. The designed 12 cm RCF is pumped by 70 fs 120 kW secant hyperbolic pulses at 1300 nm, 1400 nm and 1500 nm, respectively. One can note that the spectrum obtained by a pump centered at 1400 nm is wider than that at 1300 nm. It reaches more than two-octave spectral broadening, and it is flatter than that at 1500 nm. Furthermore, the output spectrum broadens more widely in the long wavelength range when the pump central wavelength gradually increases. Its spectral power gradually decreases below 1150 nm, which is due to the large normal dispersion for the wavelength shorter than the zero-dispersion wavelength around 1150 nm. This region has a greater requirement for the pump power, and as the pump center wavelength gradually moves away, the power spectral density generated in this region also decreases. Further research found that when the peak power of the input pulse continues to increase, the problem of insufficient pumping at 1150 nm could be solved. In simulations, we found that the spectrum can be easily extended to the short wavelength range as the nonlinear coefficient increases while wavelength decreases. Therefore, pumping at a larger wavelength can further extend the SC into the larger wavelength region. However, the overall quality of the generated SC becomes worse as the pump wavelength moves away from the local minimum of chromatic dispersion.

Then, a 1400 nm 70 fs secant hyperbolic pulse with different input peak power (P_0) is launched into a 12 cm RCF. The simulation results are shown in Figure 6b. The fiber is pumped by input pulses with different peak powers of 90 kW, 120 kW and 150 kW, respectively. It can clearly be seen that higher peak power will lead to larger SC broadening, while when it reaches a certain range, higher input peak power will induce spectral fluctuation and degrade the SC flatness. An input pulse with a peak power of 120 kW is chosen to achieve a balance between spectral broadness and its flatness. Moreover, we find that symmetrical broadening occurs at lower pump peak power resulting from SPM; nevertheless, the spectrum tends to broaden more toward the longer wavelength region at higher pump peak power as four-wave mixing (FWM) rises in the long wavelength region.

Finally, the impact of the full-width at half maximum (T_{FWHM}) of the input pulse is illustrated in Figure 6c. Short pulses of 120 kW with the central wavelength of 1400 nm possessing different pulse widths of 40 fs, 70 fs, 100 fs, respectively, are chosen as the input source launched into the designed RCF. When the peak power is a determined value, the nonlinear effect of the input pulse with a larger pump pulse width is more evident, as it contains more energy and a narrower frequency spectrum, which is the primary reason for the spectral roughness. Thus, we can obviously see that there is more fluctuation in the output SC for an input pulse with a larger pulse width. Furthermore, the output spectrum of the 40 fs input pump pulse is narrower than the others, and the power at 1150 nm is already below -40 dB. This is because the input pulse with shorter duration has a lower total input power and a wider frequency spectrum, which is equivalent to a smaller power spectral density, and thus, it does not provide enough power in the long wavelength region and at around 1150 nm. According to the additional simulation results, the pump power can be boosted to further extend the output SC span for the 40 fs input pulse case.

We finally chose 120 kW 70 fs pulse at 1400 nm as the pump source, as the corresponding femtosecond laser pump sources are already available [49]. Figure 7 illustrates the process of SC broadening using the 120 kW 70 fs input pulse with central wavelength of 1400 nm after the propagation of 0, 0.5, 1, 3, 6 and 12 cm fiber lengths, respectively. On account of the strong nonlinearity and low dispersion of the designed RCF, the generation of the broadband SC can be achieved in only a few centimeters propagation length. The nonlinear length (L_{NL}) and the dispersion length (L_D) provide the length scales over which nonlinear or dispersive effects become important for pulse evolution. These two factors can be expressed as [39]

$$L_{NL} = \frac{1}{\gamma P_0}, L_D = \frac{T_0{}^2}{|\beta_2|} \tag{4}$$

where γ is the nonlinear coefficient, P_0 and T_0 are the peak power and initial width of input pulse, respectively, β_2 is the second-order GVD. The L_{NL} and L_D for the designed RCF are

0.78 mm and 53 mm, respectively, which means nonlinearity and dispersion act together as the pulse propagates along the 12 cm long RCF. Obviously, the nonlinear effect is stronger than the dispersion effect and plays a leading role in the process of spectrum broadening.

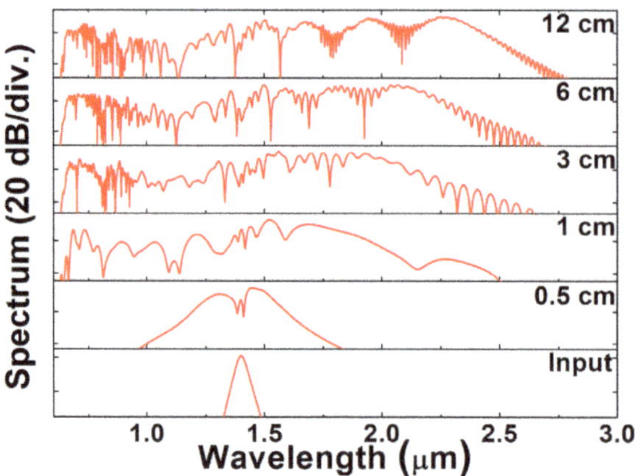

Figure 7. Broadening of the output spectra obtained at different lengths of the designed Ge-doped silica RCF.

Figure 8a displays the temporal evolutions of the corresponding SC under different propagation distances, while Figure 8b illustrates the spectral evolutions. After propagating through a 12 cm designed RCF, the SC is extended to approximately 2600 nm wavelength range spanning over two octaves. In the first place, the output pulse is symmetrically broadened around the input pulse wavelength because of the SPM. In the anomalous dispersion region, the high-order soliton effect widens the SC spectrum in the frequency domain and compresses it in the time domain. Then, it splits into fundamental-order solitons, which evolve into several impulse components in the time domain. The broadened spectrum generates dispersive waves in the normal dispersion region. Then, in the normal dispersion region, optical wave breaking (OWB) leads to the generation of new spectrum components [50], and finally, a SC is obtained. The walk-off effect gradually becomes apparent after propagating 1 cm, as illustrated in Figure 8b, which results from the accumulated dispersion. After 8 cm propagation length, the spectrum broadening tends to be stable and smoother, as shown in Figure 8a.

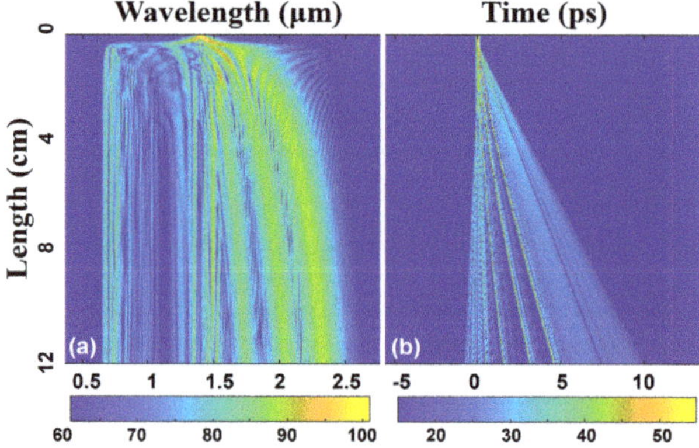

Figure 8. (a) **Spectral** and (b) temporal evolutions of the pump pulses with $OAM_{1,1}$ mode propagating along the 12 cm optimized RCF.

5. Conclusions and Perspective

In summary, we design a 50 mol% Ge-doped ring-core silica fiber supporting OAM modes with optimized parameters of 1 μm inner SiO_2 radius and 1.8 μm Ge-doped ring width, making the ring-core fiber possessing flat and low dispersion with a small variation of $<\pm30$ ps/nm/km from 1040 to 2810 nm. It is used for OAM SC generation with a 70 fs 120 kW secant hyperbolic pump pulse centered at 1400 nm. The simulation results show that an over-two-octave SC carrying the $OAM_{1,1}$ mode in the near-infrared region can be generated after 12 cm long RCF, expanding from 630 nm to 2760 nm. This designed RCF can well support the propagation of OAM modes and generate the corresponding SC, which could be utilized for various optical applications. Moreover, it is also prospective that SC carrying higher-order OAM modes can be potentially achieved with the RCF.

Author Contributions: Conceptualization, Y.Y.; data curation, J.Y.; formal analysis, J.Y., Y.W., Y.F., W.G. and Y.Y.; funding acquisition, Y.Y.; investigation, J.Y., Y.W., Y.F. and Y.Y.; methodology, J.Y., Y.W., W.Z. and Y.Y.; project administration, Y.Y.; supervision, Y.Y.; validation, Y.W., Y.F., W.G., W.Z., C.B. and Y.Y.; visualization, J.Y.; writing—original draft preparation, J.Y. and Y.Y.; writing—review and editing, Y.W., W.G., C.B., Y.R., Z.W., Y.L., Z.P. and Y.Y. All authors have read and agreed to the published version of the manuscript.

Funding: This work was supported by National Key Research and Development Program of China (2019YFB1803700); Key Technologies R & D Program of Tianjin (20YFZCGX00440); Shaanxi Key Laboratory of Deep Space Exploration Intelligent Information Technology under Grant (2021SYS-04).

Institutional Review Board Statement: Not applicable.

Informed Consent Statement: Not applicable.

Data Availability Statement: Data underlying the results presented in this paper are not available to the public but can be obtained from the authors upon reasonable request.

Conflicts of Interest: The authors declare no conflict of interest.

References

1. Bozinovic, N.; Yue, Y.; Ren, Y.; Tur, M.; Kristensen, P.; Huang, H.; Willner, A.E.; Ramachandran, S. Terabit-Scale Orbital Angular Momentum Mode Division Multiplexing in Fibers. *Science* **2013**, *340*, 1545–1548. [CrossRef] [PubMed]
2. Wang, J.; Yang, J.-Y.; Fazal, I.M.; Ahmed, N.; Yan, Y.; Huang, H.; Ren, Y.; Yue, Y.; Dolinar, S.; Tur, M.; et al. Terabit free-space data transmission employing orbital angular momentum multiplexing. *Nat. Photonics* **2012**, *6*, 488–496. [CrossRef]
3. Yan, Y.; Xie, G.; Lavery, M.; Huang, H.; Ahmed, N.; Bao, C.; Ren, Y.; Cao, Y.; Li, L.; Zhao, Z.; et al. High-capacity millimetre-wave communications with orbital angular momentum multiplexing. *Nat. Commun.* **2014**, *5*, 4876. [CrossRef] [PubMed]
4. Djordjevic, I.B. Deep-space and near-Earth optical communications by coded orbital angular momentum (OAM) modulation. *Opt. Express* **2011**, *19*, 14277–14289. [CrossRef]
5. Deng, D.; Li, Y.; Zhao, H.; Han, Y.; Ye, J.; Qu, S. High-capacity spatial-division multiplexing with orbital angular momentum based on multi-ring fiber. *J. Opt.* **2019**, *21*, 055601. [CrossRef]
6. Li, J.; Zhang, M.; Wang, D.; Wu, S.; Zhan, Y. Joint atmospheric turbulence detection and adaptive demodulation technique using the CNN for the OAM-FSO communication. *Opt. Express* **2018**, *26*, 10494–10508. [CrossRef]
7. Yu, W.; Ji, Z.; Dong, D.; Yang, X.; Xiao, Y.; Gong, Q.; Xi, P.; Shi, K. Super-resolution deep imaging with hollow Bessel beam STED microscopy. *Laser Photonics Rev.* **2015**, *10*, 147–152. [CrossRef]
8. Yan, L.; Kristensen, P.; Ramachandran, S. Vortex fibers for STED microscopy. *APL Photonics* **2019**, *4*, 022903. [CrossRef]
9. Niederriter, R.D.; Siemens, M.E.; Gopinath, J.T. Fiber Optic Sensors Based on Orbital Angular Momentum. In Proceedings of the 2015 Conference on Lasers and Electro-Optics (CLEO), San Jose, CA, USA, 10–15 May 2015.
10. Xie, G.; Song, H.; Zhao, Z.; Milione, G.; Ren, Y.; Liu, C.; Zhang, R.; Bao, C.; Li, L.; Wang, Z.; et al. Using a complex optical orbital-angular-momentum spectrum to measure object parameters. *Opt. Lett.* **2017**, *42*, 4482–4485. [CrossRef]
11. Liu, K.; Cheng, Y.; Li, X.; Gao, Y. Microwave-Sensing Technology Using Orbital Angular Momentum: Overview of Its Advantages. *IEEE Veh. Technol. Mag.* **2019**, *14*, 112–118. [CrossRef]
12. Knöner, G.; Parkin, S.; Nieminen, T.A.; Loke, V.L.Y.; Heckenberg, N.R.; Rubinsztein-Dunlop, H. Integrated optomechanical microelements. *Opt. Express* **2007**, *15*, 5521–5530. [CrossRef]
13. Duocastella, M.; Arnold, C. Bessel and annular beams for materials processing. *Laser Photonics Rev.* **2012**, *6*, 607–621. [CrossRef]
14. Liu, K.; Cheng, Y.; Gao, Y.; Li, X.; Qin, Y.; Wang, H. Super-resolution radar imaging based on experimental OAM beams. *Appl. Phys. Lett.* **2017**, *110*, 164102. [CrossRef]

15. Swartzlander, G.A.; Ford, E.L.; Abdul-Malik, R.S.; Close, L.M.; Peters, M.A.; Palacios, D.M.; Wilson, D.W.; Swartzlander, J.G.A. Astronomical demonstration of an optical vortex coronagraph. *Opt. Express* **2008**, *16*, 10200–10207. [CrossRef] [PubMed]
16. Liu, H.; Liu, K.; Cheng, Y.; Wang, H. Microwave Vortex Imaging Based on Dual Coupled OAM Beams. *IEEE Sens. J.* **2019**, *20*, 806–815. [CrossRef]
17. Liu, H.; Wang, Y.; Wang, J.; Liu, K.; Wang, H. Electromagnetic Vortex Enhanced Imaging Using Fractional OAM Beams. *IEEE Antennas Wirel. Propag. Lett.* **2021**, *20*, 948–952. [CrossRef]
18. Wang, X.; Nie, Z.; Liang, Y.; Wang, J.; Li, T.; Jia, B. Recent advances on optical vortex generation. *Nanophotonics* **2018**, *7*, 1533–1556. [CrossRef]
19. Zhu, L.; Wang, J. A review of multiple optical vortices generation: Methods and applications. *Front. Optoelectron.* **2019**, *12*, 52–68. [CrossRef]
20. Yue, Y.; Yan, Y.; Ahmed, N.; Yang, J.Y.; Zhang, L.; Ren, Y.; Huang, H.; Birnbaum, K.M.; Erkmen, B.I.; Dolinar, S.; et al. Mode Properties and Propagation Effects of Optical Orbital Angular Momentum (OAM) Modes in a Ring Fiber. *IEEE Photon. J.* **2012**, *4*, 535–543.
21. Rumala, Y.S.; Milione, G.; Nguyen, T.A.; Pratavieira, S.; Hossain, Z.; Nolan, D.; Slussarenko, S.; Karimi, E.; Marrucci, L.; Alfano, R.R. Tunable supercontinuum light vector vortex beam generator using a q-plate. *Opt. Lett.* **2013**, *38*, 5083–5086. [CrossRef]
22. Wu, R.; Torres-Company, V.; Leaird, D.E.; Weiner, A.M. Supercontinuum-based 10-GHz flat-topped optical frequency comb generation. *Opt. Express* **2013**, *21*, 6045–6052. [CrossRef] [PubMed]
23. Lamb, E.S.; Carlson, D.R.; Hickstein, D.D.; Stone, J.R.; Diddams, S.A.; Papp, S.B. Optical-Frequency Measurements with a Kerr Microcomb and Photonic-Chip Supercontinuum. *Phys. Rev. Appl.* **2018**, *9*, 024030. [CrossRef]
24. Humbert, G.; Wadsworth, W.J.; Leon-Saval, S.G.; Knight, J.C.; Birks, T.A.; Russell, P.S.J.; Lederer, M.J.; Kopf, D.; Wiesauer, K.; Breuer, E.I.; et al. Supercontinuum generation system for optical coherence tomography based on tapered photonic crystal fibre. *Opt. Express* **2006**, *14*, 1596–1603. [CrossRef] [PubMed]
25. Ferhat, M.L.; Cherbi, L.; Haddouche, I. Supercontinuum generation in silica photonic crystal fiber at 1.3 µm and 1.65 µm wavelengths for optical coherence tomography. *Optik* **2018**, *152*, 106–115. [CrossRef]
26. Morioka, T.; Mori, K.; Saruwatari, M. More than 100-wavelength-channel picosecond optical pulse generation from single laser source using supercontinuum in optical fibres. *Electron. Lett.* **1993**, *29*, 862–864. [CrossRef]
27. Oda, S.; Maruta, A. A novel quantization scheme by slicing supercontinuum spectrum for all-optical analog-to-digital conversion. *IEEE Photon. Technol. Lett.* **2005**, *17*, 465–467. [CrossRef]
28. Sotobayashi, H.; Chujo, W.; Kitayama, K.-I. Photonic gateway: TDM-to-WDM-to-TDM conversion and reconversion at 40 Gbit/s (4 channels × 10 Gbits/s). *J. Opt. Soc. Am. B* **2002**, *19*, 2810–2816. [CrossRef]
29. Neshev, D.N.; Dreischuh, A.; Maleshkov, G.; Samoc, M.; Kivshar, Y.S. Supercontinuum generation with optical vortices. *Opt. Express* **2010**, *18*, 18368–18373. [CrossRef]
30. Prabhakar, G.; Gregg, P.; Rishoj, L.; Kristensen, P.; Ramachandran, S. Octave-wide supercontinuum generation of light-carrying orbital angular momentum. *Opt. Express* **2019**, *27*, 11547–11556. [CrossRef]
31. Yue, Y.; Zhang, L.; Yan, Y.; Ahmed, N.; Yang, J.-Y.; Huang, H.; Ren, Y.; Dolinar, S.; Tur, M.; Willner, A.E. Octave-spanning supercontinuum generation of vortices in an As_2S_3 ring photonic crystal fiber. *Opt. Lett.* **2012**, *37*, 1889–1891. [CrossRef]
32. Wang, Y.; Fang, Y.; Geng, W.; Jiang, J.; Wang, Z.; Zhang, H.; Bao, C.; Huang, H.; Ren, Y.; Pan, Z.; et al. Beyond Two-Octave Coherent OAM Supercontinuum Generation in Air-Core As_2S_3 Ring Fiber. *IEEE Access* **2020**, *8*, 96543–96549. [CrossRef]
33. Geng, W.; Bao, C.; Fang, Y.; Wang, Y.; Li, Y.; Wang, Z.; Liu, Y.-G.; Huang, H.; Ren, Y.; Pan, Z.; et al. 1.6-Octave Coherent OAM Supercontinuum Generation in As_2S_3 Photonic Crystal Fiber. *IEEE Access* **2020**, *8*, 168177–168185. [CrossRef]
34. Wang, Y.; Bao, C.; Jiang, J.; Fang, Y.; Geng, W.; Wang, Z.; Zhang, W.; Huang, H.; Ren, Y.; Pan, Z. Two-octave supercontinuum generation of high-order OAM modes in air-core As_2S_3 ring fiber. *IEEE Access* **2020**, *8*, 114135–114142. [CrossRef]
35. Shiryaev, V.; Kosolapov, A.; Pryamikov, A.; Snopatin, G.; Churbanov, M.; Biriukov, A.; Kotereva, T.; Mishinov, S.; Alagashev, G.; Kolyadin, A.N. Development of technique for preparation of As_2S_3 glass preforms for hollow core microstructured optical fibers. *J. Optoelectron. Adv. Mater.* **2014**, *16*, 1020–1025.
36. Sakaguchi, S.; Todoroki, S.-I. Optical properties of GeO_2 glass and optical fibers. *Appl. Opt.* **1997**, *36*, 6809–6814. [CrossRef]
37. Dianov, E.; Mashinsky, V.; Neustruev, V.; Sazhin, O.; Guryanov, A.; Khopin, V.; Vechkanov, N.; Lavrishchev, S. Origin of Excess Loss in Single-Mode Optical Fibers with High GeO_2-Doped Silica Core. *Opt. Fiber Technol.* **1997**, *3*, 77–86. [CrossRef]
38. Brunet, C.; Ung, B.; Wang, L.; Messaddeq, Y.; LaRochelle, S.; Rusch, L.A. Design of a family of ring-core fibers for OAM transmission studies. *Opt. Express* **2015**, *23*, 10553–10563. [CrossRef]
39. Agrawal, G.P. *Nonlinear Fiber Optics*, 5th ed.; Springer: Berlin/Heidelberg, Germany, 2000; pp. 195–211.
40. Wang, C.C.; Wang, M.H.; Wu, J. Heavily Germanium-Doped Silica Fiber with a Flat Normal Dispersion Profile. *IEEE Photon. J.* **2015**, *7*, 1–10. [CrossRef]
41. Dianov, E.; Mashinsky, V. Germania-based core optical fibers. *J. Light. Technol.* **2005**, *23*, 3500–3508. [CrossRef]
42. Lægsgaard, J.; Tu, H. How long wavelengths can one extract from silica-core fibers? *Opt. Lett.* **2013**, *38*, 4518–4521. [CrossRef]
43. Rottwitt, K.; Povlsen, J.H. Analyzing the fundamental properties of Raman amplification in optical fibers. *J. Light. Technol.* **2005**, *23*, 3597–3605. [CrossRef]
44. Yatsenko, Y.; Mavritsky, A. D-scan measurement of nonlinear refractive index in fibers heavily doped with GeO_2. *Opt. Lett.* **2007**, *32*, 3257–3259. [CrossRef] [PubMed]

45. Zhang, L.; Agarwal, A.M.; Kimerling, L.C.; Michel, J. Nonlinear Group IV photonics based on silicon and germanium: From near-infrared to mid-infrared. *Nanophotonics* **2014**, *3*, 247–268. [CrossRef]
46. Afshar, S.; Monro, T.M. Monro. A full vectorial model for pulse propagation in emerging waveguides with subwavelength structures part I: Kerr nonlinearity. *Opt. Express* **2009**, *17*, 2298–2318. [CrossRef]
47. Malitson, I.H. Interspecimen comparison of the refractive index of fused silica. *J. Opt. Soc. Am.* **1965**, *55*, 1205–1209. [CrossRef]
48. Barnes, N.P.; Piltch, M.S. Temperature-dependent Sellmeier coefficients and nonlinear optics average power limit for germanium. *J. Opt. Soc. Am.* **1979**, *69*, 178–180. [CrossRef]
49. Chen, B.; Hong, L.; Hu, C.; Li, Z. White Laser Realized via Synergic Second- and Third-Order Nonlinearities. *Research* **2021**, *2021*, 1539730. [CrossRef]
50. Heidt, A.M.; Hartung, A.; Bosman, G.; Krok, P.; Rohwer, E.; Schwoerer, H.; Bartelt, H. Coherent octave spanning near-infrared and visible supercontinuum generation in all-normal dispersion photonic crystal fibers. *Opt. Express* **2011**, *19*, 3775–3787. [CrossRef] [PubMed]

Article

Bandwidth Extension in a Mid-Link Optical Phase Conjugation

Paweł Rosa [1,*], Giuseppe Rizzelli Martella [2] and Mingming Tan [3]

[1] National Institute of Telecommunications, Szachowa 1, 04-894 Warsaw, Poland
[2] LINKS Foundation, Via Piercarlo Boggio 61, 10138 Torino, Italy
[3] Aston Institute of Photonics Technologies, Aston University, Birmingham B4 7ET, UK
* Correspondence: p.rosa@il-pib.pl

Abstract: In this paper, we investigate various designs of distributed Raman amplifier (DRA) to extend amplification bandwidth in mid-link optical phase conjugation (OPC) systems and compare bands 191–197 THz and 192–198 THz giving a total bandwidth of 6 THz using a single wavelength pump. We demonstrate the use of highly reflective fiber Bragg grating (FBG) to minimize gain variation across a WDM grid by optimizing forward and backward pump powers as well as the wavelength of FBGs for original and conjugated channels. Finally, we also simulate OSNR and Kerr nonlinear reduction as a product of signals asymmetry and nonlinear phase shift (NPS) for all channels.

Keywords: Raman amplification; optical fiber communications; optical phase conjugation

Citation: Rosa, P.; Martella, G.R.; Tan, M. Bandwidth Extension in a Mid-Link Optical Phase Conjugation. *Sensors* **2022**, *22*, 6385. https://doi.org/10.3390/s22176385

Academic Editors: Carlos Marques, Jiangbing Du, Yang Yue, Jian Zhao and Yan-ge Liu

Received: 29 June 2022
Accepted: 22 August 2022
Published: 24 August 2022

Publisher's Note: MDPI stays neutral with regard to jurisdictional claims in published maps and institutional affiliations.

1. Introduction

Optical phase conjugator (OPC) in a long-haul transmission system can effectively compensate for both linear (e.g., chromatic dispersion) and nonlinear (e.g., the fiber Kerr nonlinearity) impairments, and therefore can improve the data capacity or transmission distance [1–27]. The efficiency of how much the fiber nonlinearity can be compensated assisted by mid-link OPC was limited by several factors, such as the slope of the chromatic dispersion map of the transmission fiber and the signal power profile along the fiber [1–6]. The symmetry of chromatic dispersion slope can be tailored by optimizing the dispersion map using a combination of transmission fiber and dispersion compensating fiber [9]. This can be used together with erbium-doped fiber amplifiers (EDFA) which is the most widely used amplification technique; however, it requires special transmission fiber and dispersion compensating fiber to maintain the dispersion map symmetry. Another way to maximize fiber nonlinearity compensation efficiency with OPC is to use distributed Raman amplification to make the signal power profile symmetrical. Distributed Raman amplification generates optical gain using Raman pump lasers over the standard transmission fiber, providing distributed amplification along the whole fiber rather than a discrete or lumped amplification within a few meters of doped fiber as in EDFA. DRA can be highly flexible to specifically tailor the signal power profiles to be highly symmetrical before and after the OPC [3–6]. The pump wavelength can be adjusted with fiber Bragg gratings (FBG) with selected center wavelength [28–40]. The use of distributed Raman amplification can improve the maximum transmission distance or data capacity without the mid-link OPC (or the first half of the link before the OPC) and therefore if the symmetry of the link can be maintained at a very high level for all channels, the overall transmission performances or data capacity can be significantly improved using mid-link OPC due to the efficient compensation of fiber nonlinearity [4–6].

We show, for the first time, that the transmission bandwidth can be extended using FBGs at two different wavelengths for transmitted and conjugated channels in mid-link optical phase conjugation. The novel design approach allows for the highest OPC bandwidth using Raman amplification and gives the lowest asymmetry for a single channel up to date for designs of distributed Raman amplification schemes which are aimed to improve the symmetry of the link for the transmission systems.

2. Distributed Raman Amplification

In telecommunication systems, there are several Raman configurations that can be used. In a single span unrepeatered submarine transmission, second order bi-directional pumping with two FBGs in the front and the end of the span provides great signal power distribution along the fiber compared to first order amplification, which ultimately increases transmission reach [37–40]. Depending on the application, the same approach might not work in long-haul transmission. Forward pumping can help with signal power distribution creating a quasi-lossless transmission medium [32]; however, the benefits of low noise and signal power variation become meaningless in coherent data transmission due to forward relative intensity noise (RIN) [32–36]. In [36] we compared six different Raman configurations, including first order, second order and dual order and have shown, experimentally, that bi-directionally distributed Raman amplification with a single FBG at the end of the transmission span, forming a half-open cavity with random distributed feedback (DFB) lasing [28], significantly decreases accumulation of amplified spontaneous emission (ASE) noise built up across the transmission span and keeps signal power variation low [28–40], extending data transmission by almost 900 km to a record distance of 7915 km. In this particular scheme the forward Raman laser pump at 1366 nm amplifies the backward propagating random DFB lasing at the frequency specified by the wavelength of the FBG. This approach allows for a RIN transfer reduction [32–36] from the noisy forward Raman pump to the Stokes-shifted light, becoming an efficient solution for a long-haul coherent data transmission format [33–36].

The schematic design of the random DFB Raman laser amplifier is shown in Figure 1. Two Raman fiber laser pumps downshifted in wavelength to 1366 nm (approximately two Stokes shifts from the signal) were located at each end of the standard single mode fiber (SMF). The span length was 60 km. A half-open cavity random DFB laser was formed at the wavelength of the FBG at the end of the fiber that amplifies original and conjugated WDM channels in the OPC system. We assume that fiber used for the FBGs is the same as transmission fiber, which is standard SMF fiber. By optimizing the wavelength of the FBG, rather than deploying a seed at different wavelength, the spectral gain profile of the amplified WDM signals can be modified and enhanced. This is visualized by an example shown in Figure 2 where we compare the Raman gain shift resulting from FBGs at different wavelength. To avoid polarization gain dependance in WDM transmission, Raman pump lasers at both ends and lasing at the wavelength of the FBG were fully depolarized.

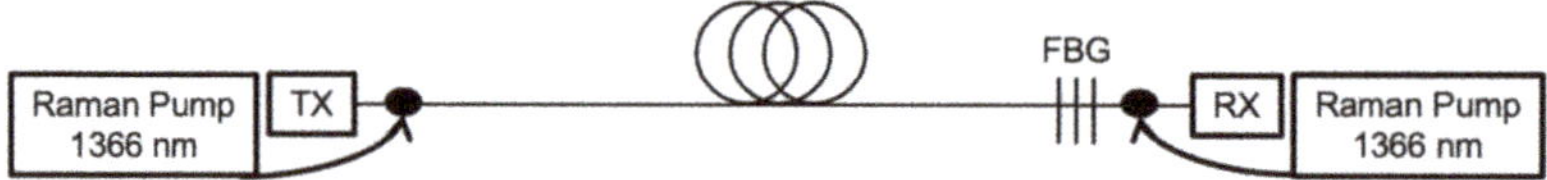

Figure 1. Raman fiber laser-based amplifier with a half-open cavity random lasing.

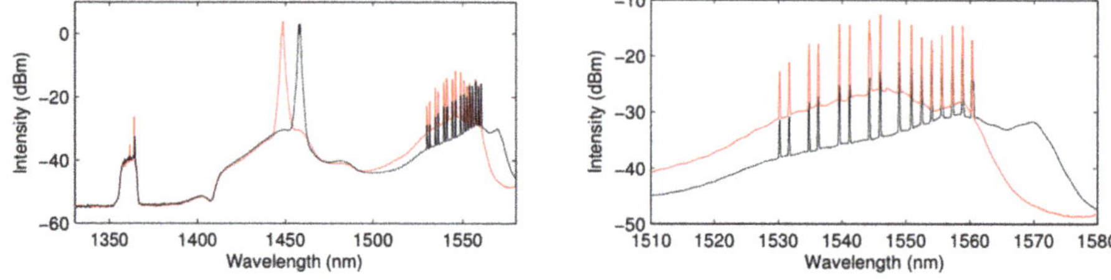

Figure 2. The Raman gain shift using FBGs centered at 1448 nm (red) and 1458 nm (black). The right figure is zoomed on the 16 WDM signals wavelength range.

3. Simulation

The spectral components of the OPC system described in Section 2 was simulated using an experimentally confirmed model [32–40] that takes into account residual Raman gain from the primary pump at 1366 nm to the signal in the C-band as well as pump depletion from both pumps to the lower order pumps and the signal components, double Rayleigh scattering (DRS) and amplified spontaneous emission (ASE) noise for each of the signals. Noise was calculated in a room temperature of 24 °C and 0.1 nm bandwidth. Wavelength dependent Raman gain coefficient and attenuation factor were independently chosen for each WDM component as well as primary pump and lasing at the frequency of the FBG using tables for the SMF. Rayleigh backscattering coefficients for Raman pump, lasing and signal were 1.0×10^{-4}, 6.5×10^{-5} and 4.5×10^{-5} km^{-1}, respectively. In [28] we showed that higher reflectivity (99%) provides lower nonlinear phase shift compared with 70% and 50% as well as better power efficiency performance without sacrificing output OSNR. To find the best gain profile, the center wavelength of high reflectivity (99%) FBGs was varied from 1456 to 1464 nm for WDM section of transmitted channels and from 1442 to 1454 nm for the conjugated copy with a 2 nm step. The loss and bandwidth of the FBG in the simulations was set to 0.2 dB and 200 GHz, respectively.

Figure 3 illustrates the WDM grid of the mid-link OPC system. Total bandwidth of 6 THz consists of 30 transmitted WDM channels with 100 GHz spacing ranging from 192–194.9 THz and 30 WDM conjugated copies of the original signal in the range 195.1–198 THz (we also present results for 191–197 THz grid). We assumed 200 GHz as the guardband in the middle of the grid centered at 195 THz for the optical phase conjugator. The signal power profile of each WDM component of the original and conjugated signal was simulated independently.

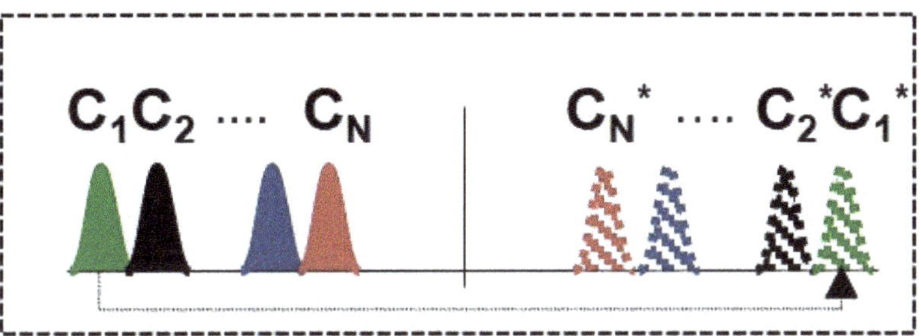

Figure 3. WDM grid of the transmitted (**left**) and the conjugated (**right**) WDM channels in a mid-link OPC system. "*" denotes conjugated channels.

Forward pump power of the Raman fibre laser at 1366 nm was simulated from 0.7 to 1.6 W with 100 mW step. Backward pump power was simulated to give 0 dB net gain for the channel under investigation, then the rest of the WDM channels were simulated with the same forward and backward pump powers: one set of results consists of 30 possible combinations (optimisation towards first channel CH1, second channel CH2 and so on). Conjugated signals were simulated with the same pocedure. To summarise, we simulated every possible combination varying forward pump power (0.7–1.6 W) for each installed FBG (1442–1454 nm for original and 1456–1464 nm for conjugated WDM grid) and finally compared asymmetry between both WDM channel sets to achieve the best overall asymmetry performance in an OPC system with second order Raman amplification.

The asymmetry was calculated using formula:

$$\frac{\int_0^L |P_1(z) - P_2(L-z)| \, dz}{\int_0^L [P_1(z) + P_2(z)]/2 \, dz} \times 100 \tag{1}$$

where L is the span length and P_1 and P_2 represent signal power evolution of the transmitted and conjugated channels, respectively.

4. Results and Discussion

4.1. 191–197 THz WDM Grid

For a given WDM grid bandwidth, the key factors that need to be optimized for the best performance overall in a mid-link OPC system are the wavelength of the FBG and the forward and backward pump powers. Both optimizations must be done separately for the original and conjugated WDM grid. Our initial investigation was done for a WDM grid 191–197 THz. Firstly, we had to select the right set of FBGs that will have a performance impact for residual channels located at the beginning and the end of the band.

In Figure 4 we show the impact of the FBG wavelength on asymmetry performance in WDM grid under investigation. Here, the central wavelength of the FBG for original channels was set to 1462 nm (the best configuration) and the one of the FBG for conjugated channels was varied from 1442–1450 nm. The asymmetry of residual side channels can be improved by 12% (CH1: 21% with 1450 nm and 33% with 1442 nm). However, this advantage is lost for CH6 and CH7 where the asymmetry performance is reversed, hence we need average overall performance. In this case the matching FBGs pair was 1462 nm for original and 1446 nm for conjugated WDM grid (we also simulated FBG for original WDM grid from 1456–1464 nm, however, for clarity we only show the best performing configuration with 1462 nm). Next, we optimized forward and backward pump powers for original and conjugated WDM grid.

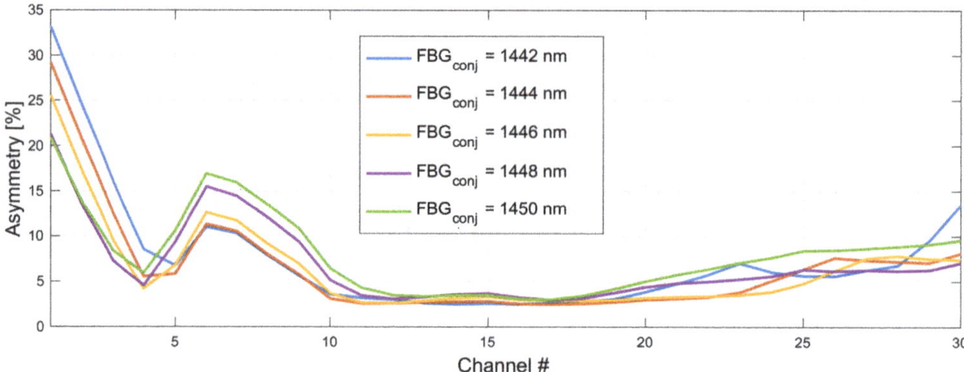

Figure 4. Impact of a FBG choice on asymmetry performance in 191–197 THz grid. Wavelength of a FBG for original WDM grid was set to 1462 nm and FBG for the conjugated copy was varied from 1442–1450 nm.

In Figure 5 we show the impact of pump power on asymmetry performance for original and conjugated WDM grid. Forward pump power for both configurations was varied from 1–1.6 W. Backward pump power from 2.6–2.9 W for original (191–193.9 THz) and 1.6–1.9 W for conjugated (194.1–197 THz) WDM grid. The difference in backward pump power for each WDM grid results from wavelength dependent Raman gain and fiber attenuation.

By analyzing all possible combinations of FBG, forward and backward pump powers for original and conjugated WDM grid, we found the profile that gave the best average asymmetry of 6.4%; however, this figure is biased by first residual channels that are on the edge of Raman gain profile.

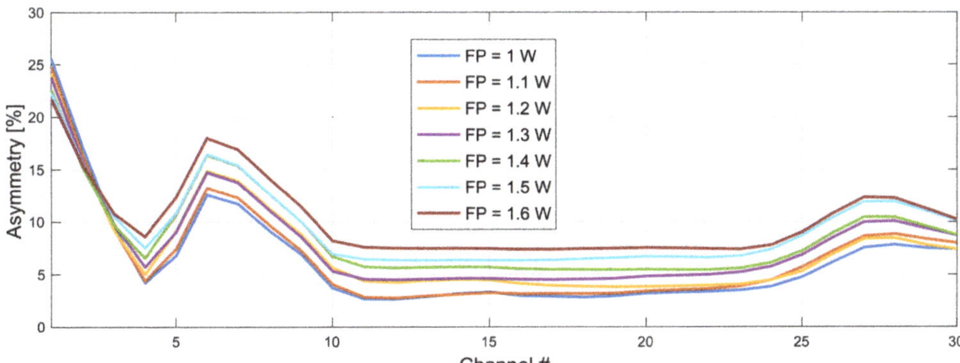

Figure 5. Impact of pump power choice (original WDM grid) on asymmetry performance. Forward pump power of the conjugated grid fixed to be 1.3 W. FBG for original and conjugated WDM grid was set to 1462 and 1446 nm, respectively.

In Figure 6 we show the best profile with the following settings: FBG 1462 nm with forward pump power 1 W for the original, and FBG 1446 nm with 1.3 W forward pump power for the conjugated WDM grid.

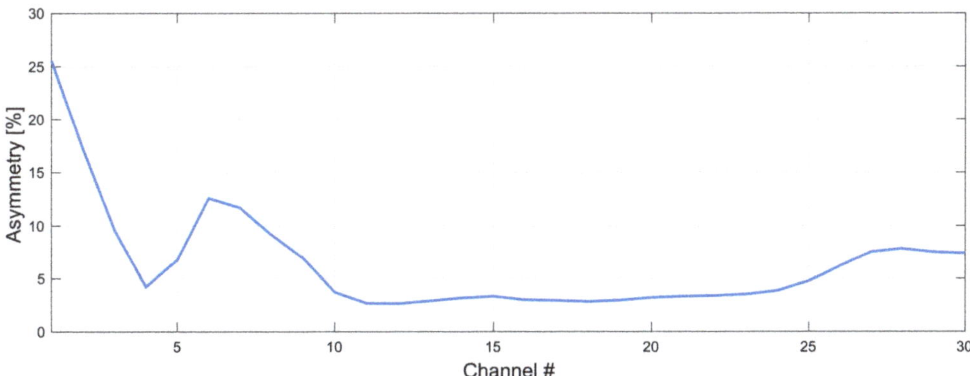

Figure 6. WDM grid with the best average asymmetry of 6.4%. Forward pump power was set to be 1 W and 1.3 W and FBG to 1462 nm and 1446 nm for original and conjugated WDM grid, respectively.

4.2. 192–198 THz WDM Grid

Based on the results shown in Section 4.1, we decided to repeat the simulations with a grid shifted by 1 THz to avoid the worst performing channels in the beginning of the spectrum. The new grid was set to 192–198 THz. Based on the knowledge from the previous configuration we only tested the FBG sets 1458–1460 nm and 1450–1454 nm for original and conjugated WDM grid, respectively.

In Figures 7 and 8, we show the impact of a FBG choice on asymmetry performance in 192–198 THz grid. The optimum configuration giving the lowest average asymmetry was with FBG set to 1460 nm and 1450 nm for the original (Figure 9) and conjugated (Figure 10) WDM grid, respectively. The right choice of a FBG helps to optimize the gain profile of the WDM band, which will particularly have an impact for an on–off gain of residual front and end channels. This will directly affect asymmetry performance.

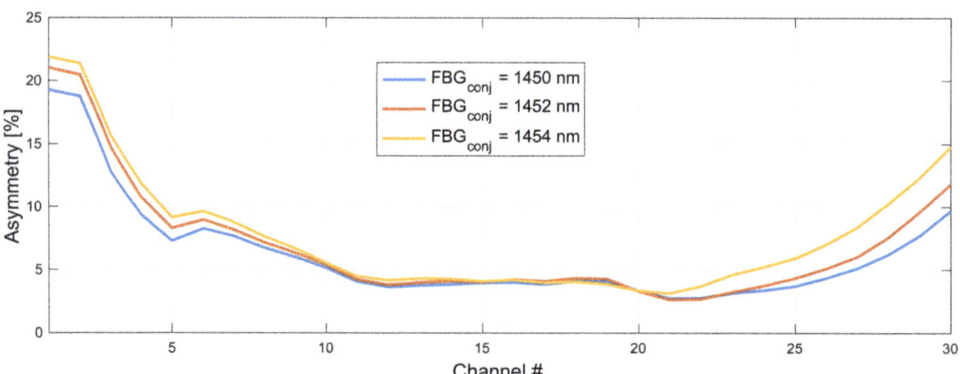

Figure 7. Impact of a FBG choice on asymmetry performance in 192–198 THz grid. FBG for original WDM grid was set to 1458 nm and conjugated varied from 1450–1454 nm.

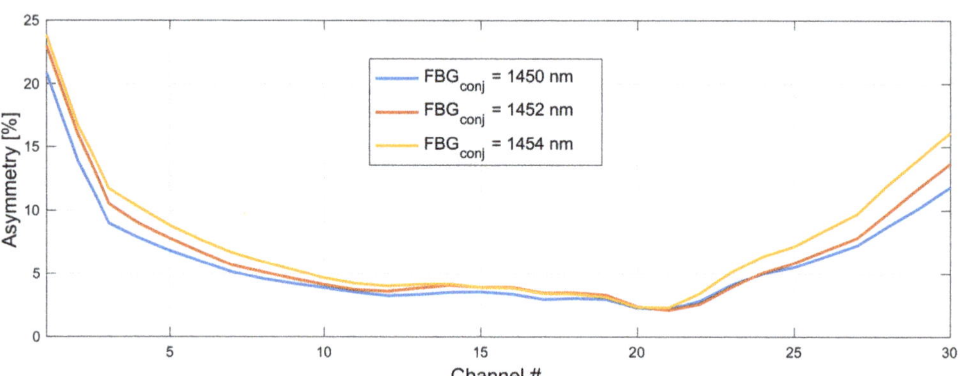

Figure 8. Impact of a FBG choice on asymmetry performance in 192–198 THz grid. FBG for original WDM grid was set to 1460 nm and conjugated varied from 1450–1454 nm.

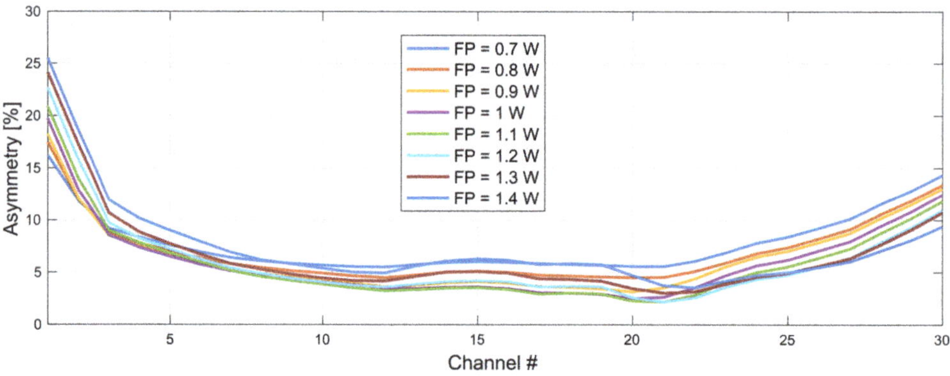

Figure 9. Impact of pump power choice (original WDM grid) on asymmetry performance. Forward pump power of the conjugated grid was fixed to be 1.3 W. FBG for original and conjugated WDM grid was set to 1460 and 1450 nm, respectively.

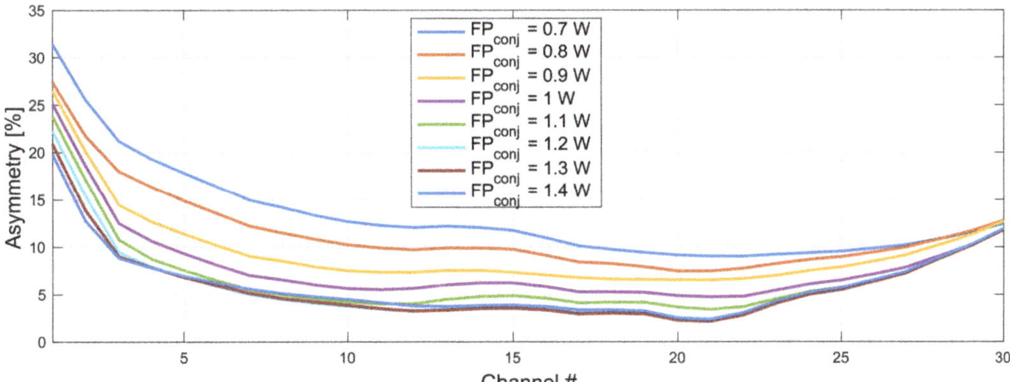

Figure 10. Impact of pump power choice (conjugated WDM grid) on asymmetry performance. Forward pump power of the original grid was fixed to be 1.1 W. FBG for original and conjugated WDM grid was set to 1460 and 1450 nm, respectively.

After choosing the right FBG set for our WDM band we optimized forward and backward pump powers. This time we extended the range and simulated 0.7–1.4 W forward pump power. Backward pump power was stable for all possible configurations and oscillated within the 50 mW range around 2.3 W for the original and 1.8 W for the conjugated WDM grid.

By shifting our band by 1 THz we managed to improve average asymmetry by 0.5% achieving 5.9% with only three channels (CH1, CH2 and CH30) with figures above 10%. In this regime (asymmetry below 10%) we exceeded 35 nm bandwidth of the C band (1530–1565 nm) from ~4.3 THz to 5.4 THz (27 WDM channels with asymmetry below 10%, giving total 54 channels). As seen in Figure 11, the majority of channels (CH7–CH24) were below 5%, which is a very impressive result covering 3.4 THz.

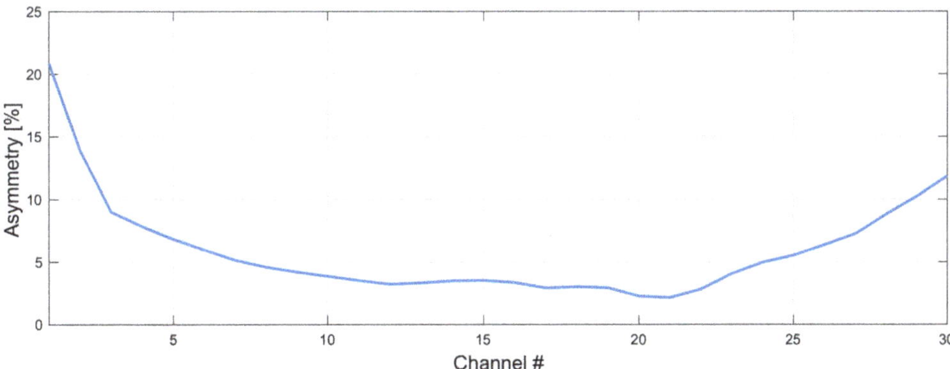

Figure 11. WDM grid with the best average asymmetry of 5.9%. Forward pump power was set to be 1.1 W and 1.3 W and FBG to 1460 nm and 1450 nm for original and conjugated WDM grid, respectively.

In a single channel regime CH21 achieved asymmetry 2.1%, which is the lowest asymmetry in a 60 km span up to date (see Figure 12). This value most likely would be even lower if a single channel only would be simulated with the same configuration due to the Raman pump depletion. Signal power variation in the original and conjugated channels is very low, below 2 dB, comparing to 12 dB when using EDFA.

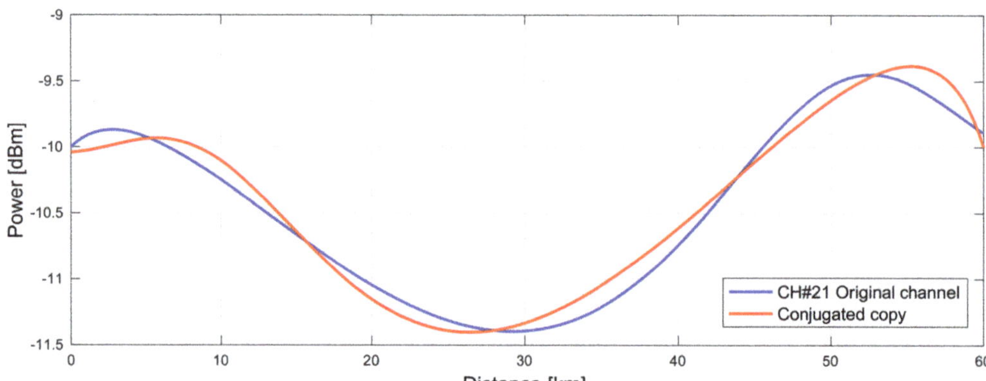

Figure 12. Signal power profile of original (blue) and conjugated (red) channel with lowest asymmetry of 2.1% in a 60 km standard SMF span.

In Figure 13 we show OSNR and NPS results for an optimized 6 THz WDM grid (192–198 THz) for both original (CH1–CH30) and conjugated (CH31–CH60) channels in a 60 km span. We can notice small OSNR variation within 1 dB range between the best and the worst performing channels for both simulated grids, thanks to the combined gain from the first order pump at 1366 nm and lasing at the wavelength of the FBG. NPS variation was also very low across all but the first two channels; however, at these values of NPS it is insignificant.

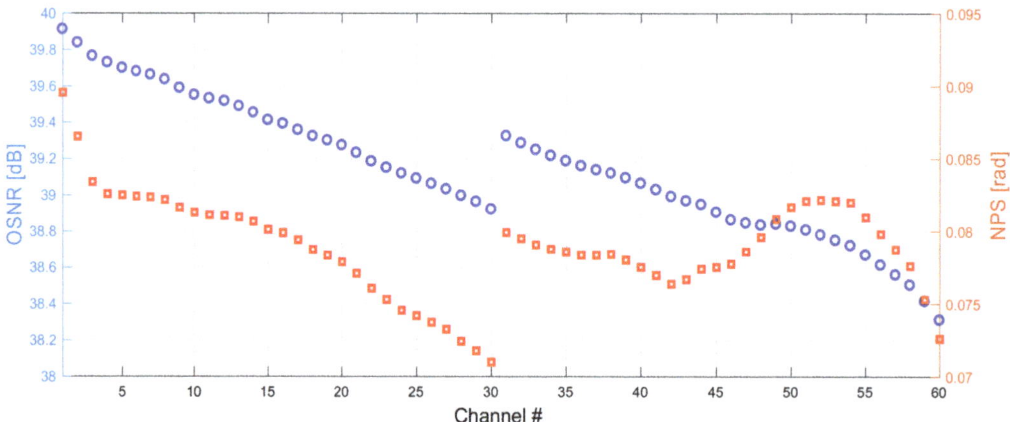

Figure 13. OSNR (blue) and NPS (red) simulations for original (CH1–CH30) and conjugated (CH31–CH60) channels.

5. Conclusions

We present, for the first time, the use of half-open cavity random DFB Raman laser amplification with combinations of different FBGs to extend the bandwidth of the mid-link OPC system beyond the standard C band. We performed gain profile optimization in a 60 km standard SMF span utilizing a 6 THz bandwidth with total of 60 WDM channels. With a fine optimization of FBGs and pump powers, we achieved for the first time a record

5.9% average asymmetry in a 192–198 THz band using single wavelength 1366 nm Raman laser pump only in an OPC system. We also achieved 2.1% asymmetry performance for a single channel, which is the lowest asymmetry achieved to date in a 60 km standard SMF span length. The results show the potential for further improvement by utilizing 6 THz band using DWDM with a 25 GHz or even 12.5 GHz spacing that would increase the total transmission capacity of an OPC system by an order of magnitude.

Author Contributions: P.R., G.R.M. and M.T. proposed the concept and initiated the study. P.R. and G.R.M. carried out numerical simulations and conducted analytical calculations. M.T. supervised the studies. The paper was written by and P.R., G.R.M. and M.T. All authors have read and agreed to the published version of the manuscript.

Funding: This research was funded by a UK Engineering and Physical Sciences Research Council (EPSRC) Grant (ARGON), a Royal Society International Exchange Grant (IEC\NSFC\211244), the Aston Impact Fund, and a Polish Ministry of Science and Higher Education Grant (12300051).

Institutional Review Board Statement: Not applicable.

Informed Consent Statement: Not applicable.

Data Availability Statement: Original data is available at Aston Research Explorer (https://doi.org/10.17036/researchdata.aston.ac.uk.00000575).

Acknowledgments: Not applicable.

Conflicts of Interest: The authors declare no conflict of interest.

References

1. Jasen, S.L.; van den Borne, D.; Spinnler, B.; Calabro, S.; Suche, H.; Krummrich, P.M.; Sohler, W.; Khoe, G.-D. Optical phase conjugation for ultra long-haul phase-shift-keyed transmission. *J. Light. Technol.* **2006**, *24*, 54–64. [CrossRef]
2. Minzioni, P.; Cristiani, I.; Degiorgio, V.; Marazzi, L.; Martinelli, M.; Langrock, C.; Fejer, M.M. Experimental Demonstration of Nonlinearity and Dispersion Compensation in an Embedded Link by Optical Phase Conjugation. *IEEE Photonics Technol. Lett.* **2006**, *18*, 995–997. [CrossRef]
3. Tan, M.; Rosa, P.; Nguyen, T.T.; Al-Khateeb, M.A.Z.; Iqbal, M.A.; Xu, T.; Wen, F.; Ania-Castañón, J.D.; Ellis, A.D. Distributed Raman Amplification for Fiber Nonlinearity Compensation in a Mid-Link Optical Phase Conjugation System. *Sensors* **2022**, *22*, 758. [CrossRef]
4. Tan, M.; Nguyen, T.T.; Rosa, P.; Al-Khateeb, M.A.Z.; Zhang, T.T.; Ellis, A.D. Enhancing the Signal Power Symmetry for Optical Phase Conjugation Using Erbium-Doped-Fiber-Assisted Raman Amplification. *IEEE Access* **2020**, *8*, 222766–222773. [CrossRef]
5. Rosa, P.; Rizzelli, G.; Ania-Castañón, J.D. Link optimization for DWDM transmission with an optical phase conjugation. *Opt. Express* **2016**, *24*, 16450–16455. [CrossRef] [PubMed]
6. Al-Khateeb, M.; Tan, M.; Zhang, T.; Ellis, A. Combating Fiber Nonlinearity Using Dual-Order Raman Amplification and OPC. *IEEE Photonics Technol. Lett.* **2019**, *31*, 877–880. [CrossRef]
7. Du, L.B.; Morshed, M.M.; Lowery, A.J. Fiber nonlinearity compensation for OFDM super-channels using optical phase conjugation. *Opt. Express* **2012**, *20*, 19921. [CrossRef]
8. Huang, C.; Shu, C. Raman-enhanced optical phase conjugator in WDM transmission systems. *Opt. Express* **2018**, *26*, 10274. [CrossRef]
9. Kaminski, P.; Da Ros, F.; Clausen, A.T.; Forchhammer, S.; Oxenløwe, L.K.; Galili, M. Improved nonlinearity compensation of OPC-aided EDFA-amplified transmission by enhanced dispersion mapping. In Proceedings of the 2020 Conference on Lasers and Electro-Optics (CLEO), San Jose, CA, USA, 10–15 May 2020.
10. Yoshima, S.; Sun, Y.; Liu, Z.; Bottrill, K.R.H.; Parmigiani, F.; Richardson, D.J.; Petropoulos, P. Mitigation of Nonlinear Effects on WDM QAM Signals Enabled by Optical Phase Conjugation with Efficient Bandwidth Utilization. *J. Light. Technol.* **2017**, *35*, 971–978. [CrossRef]
11. Sackey, I.; Schmidt-Langhorst, C.; Elschner, R.; Kato, T.; Tanimura, T.; Watanabe, S.; Hoshida, T.; Schubert, C. Waveband-Shift-Free Optical Phase Conjugator for Spectrally Efficient Fiber Nonlinearity Mitigation. *J. Light. Technol.* **2018**, *36*, 1309–1317. [CrossRef]
12. Pelusi, M.D. WDM signal All-optical Precompensation of Kerr Nonlinearity in Dispersion-Managed Fibers. *IEEE Photonics Technol. Lett.* **2013**, *25*, 71–73. [CrossRef]
13. Solis-Trapala, K.; Pelusi, M.; Tan, H.N.; Inoue, T.; Namiki, S. Optimized WDM Transmission Impairment Mitigation by Multiple Phase Conjugations. *J. Light. Technol.* **2016**, *34*, 431–440. [CrossRef]
14. Ellis, A.D.; McCarthy, M.E.; Al-Khateeb, M.A.Z.; Sygletos, S. Capacity limits of systems employing multiple optical phase conjugators. *Opt. Express* **2015**, *23*, 20381–20393. [CrossRef] [PubMed]

15. Phillips, I.; Tan, M.; Stephens, M.F.; McCarthy, M.; Giacoumidis, E.; Sygletos, S.; Rosa, P.; Fabbri, S.; Le, S.T.; Kanesan, T.; et al. Exceeding the Nonlinear-Shannon Limit using Raman Laser Based Amplification and Optical Phase Conjugation. In Proceedings of the Optical Fiber Communication Conference, OSA Technical Digest, San Francisco, CA, USA, 9–13 March 2014.

16. Stephens, M.F.C.; Tan, M.; Phillips, I.D.; Sygletos, S.; Harper, P.; Doran, N.J. 1.14Tb/s DP-QPSK WDM polarization-diverse optical phase conjugation. *Opt. Express* **2014**, *22*, 11840. [CrossRef]

17. Da Ros, F.; Yankov, M.P.; Silva, E.P.d.; Pu, M.; Ottaviano, L.; Hu, H.; Semenova, E.; Forchhammer, S.; Zibar, D.; Galili, M.; et al. Characterization and Optimization of a High-Efficiency AlGaAs-On-Insulator-Based Wavelength Convertor for 64- and 256-QAM Signals. *J. Light. Technol.* **2017**, *35*, 3750–3757. [CrossRef]

18. Da Ros, F.; Edson, G.; Silva, E.P.d.; Peczek, A.; Mai, A.; Petermann, K.; Zimmermann, L.; Oxenløwe, L.K.; Galili, M. Optical Phase Conjugation in a Silicon Waveguide with Lateral p-i-n Diode for Nonlinearity Compensation. *J. Light. Technol.* **2019**, *37*, 323–329. [CrossRef]

19. Umeki, T.; Kazama, T.; Sano, A.; Shibahara, K.; Suzuki, K.; Abe, M.; Takenouchi, H.; Miyamoto, Y. Simultaneous nonlinearity mitigation in 92 × 180-Gbit/s PDM-16QAM transmission over 3840 km using PPLN-based guard-band-less optical phase conjugation. *Opt. Express* **2016**, *24*, 16945. [CrossRef]

20. Hu, H.; Jopson, R.M.; Gnauck, A.H.; Randel, S.; Chandrasekhar, S. Fiber nonlinearity mitigation of WDM-PDM QPSK/16-QAM signals using fiber-optic parametric amplifiers based multiple optical phase conjugations. *Opt. Express* **2017**, *25*, 1618. [CrossRef]

21. Namiki, S.; Solis-Trapala, K.; Tan, H.N.; Pelusi, M.; Inoue, T. Multi-Channel Cascadable Parametric Signal Processing for Wavelength Conversion and Nonlinearity Compensation. *J. Light. Technol.* **2017**, *35*, 815–823. [CrossRef]

22. Ellis, A.D.; McCarthy, M.E.; Al-Khateeb, M.A.Z.; Sorokina, M.; Doran, N.J. Performance limits in optical communications due to fiber nonlinearity. *Adv. Opt. Photonics* **2019**, *9*, 429–503. [CrossRef]

23. Bidaki, E.; Kumar, S. A Raman-pumped Dispersion and Nonlinearity Compensating Fiber For Fiber Optic Communications. *IEEE Photonics J.* **2020**, *12*, 720017. [CrossRef]

24. Al-Khateeb, M.A.Z.; Tan, M.; Iqbal, M.A.; Ali, A.; McCarthy, M.E.; Harper, P.; Ellis, A.D. Experimental demonstration of 72% reach enhancement of 3.6 Tbps optical transmission system using mid-link optical phase conjugation. *Opt. Express* **2018**, *26*, 23960–23968. [CrossRef] [PubMed]

25. Ellis, A.D.; Tan, M.; Iqbal, M.A.; Al-Khateeb, M.A.Z.; Gordienko, V.; Saavedra Mondaca, G.; Fabbri, S.; Stephens, M.F.C.; McCarthy, M.E.; Perentos, A.; et al. 4 Tb/s Transmission Reach Enhancement Using 10 × 400 Gb/s Super-Channels and Polarization Insensitive Dual Band Optical Phase Conjugation. *J. Light. Technol.* **2016**, *34*, 1717–1723. [CrossRef]

26. Al-Khateeb, M.A.Z.; Iqbal, M.A.; Tan, M.; Ali, A.; McCarthy, M.E.; Harper, P.; Ellis, A.D. Analysis of the nonlinear Kerr effects in optical transmission systems that deploy optical phase conjugation. *Opt. Express* **2018**, *26*, 3145–3160. [CrossRef]

27. Rosa, P.; Le, S.T.; Rizzelli, G.; Tan, M.; Harper, P.; Ania-Castañón, J.D. Signal power asymmetry optimisation for optical phase conjugation using Raman amplification. *Opt. Express* **2015**, *23*, 31772–31778. [CrossRef] [PubMed]

28. Rosa, P.; Rizzelli, G.; Tan, M.; Harper, P.; Ania-Castañón, J.D. Characterisation of random DFB Raman laser amplifier for WDM transmission. *Opt. Express* **2015**, *23*, 28634–28639. [CrossRef] [PubMed]

29. Tan, M.; Rosa, P.; Iqbal, M.A.; Phillips, I.D.; Nuno, J.; Ania-Castañón, J.D.; Harper, P. RIN mitigation in second-order pumped Raman fibre laser based amplification. In Proceedings of the Asia Communications and Photonics Conference, Hong Kong, 19–23 November 2015; OSA Technical Digest, paper AM2E.6. Optical Society of America: Washington, DC, USA, 2015.

30. Tan, M.; Rosa, P.; Le, S.T.; Dvoyrin, V.V.; Iqbal, M.A.; Sugavanam, S.; Turitsyn, S.K.; Harper, P. RIN mitigation and transmission performance enhancement with forward broadband pump. *IEEE Photonics Technol. Lett.* **2018**, *30*, 254–257. [CrossRef]

31. Rizzelli, G.; Iqbal, M.A.; Gallazzi, F.; Rosa, P.; Tan, M.; Ania-Castañón, J.D.; Krzczanowicz, L.; Corredera, P.; Phillips, I.; Forysiak, W.; et al. Impact of input FBG reflectivity and forward pump power on RIN transfer in ultralong Raman laser amplifiers. *Opt. Express* **2016**, *24*, 29170–29175. [CrossRef]

32. Ania-Castañón, J.D.; Karalekas, V.; Harper, P.; Turitsyn, S.K. Simultaneous spatial and spectral transparency in ultralong fiber lasers. *Phys. Rev. Lett.* **2008**, *101*, 123903. [CrossRef]

33. Tan, M.; Rosa, P.; Phillips, I.D.; Harper, P. Long-haul Transmission Performance Evaluation of Ultra-long Raman Fiber Laser Based Amplification Influenced by Second Order Co-pumping. In Proceedings of the Asia Communications and Photonics Conference, Shanghai, China, 11–14 November 2014; OSA Technical Digest (online), paper ATh1E.4. Optical Society of America: Washington, DC, USA, 2014.

34. Tan, M.; Rosa, P.; Le, S.T.; Phillips, I.D.; Harper, P. Evaluation of 100 g DP-QPSK long-haul transmission performance using second order co-pumped Raman laser based amplification. *Opt. Express* **2015**, *23*, 22181–22189. [CrossRef]

35. Tan, M.; Rosa, P.; Phillips, I.D.; Harper, P. Extended Reach of 116 Gb/s DP-QPSK Transmission using Random DFB Fiber Laser Based Raman Amplification and Bidirectional Second-order Pumping. In Proceedings of the Optical Fiber Communication Conference, OSA Technical Digest, Los Angeles, CA, USA, 22–26 March 2015.

36. Tan, M.; Rosa, P.; Le, S.T.; Iqbal, M.A.; Phillips, I.D.; Harper, P. Transmission performance improvement using random DFB laser based Raman amplification and bidirectional second-order pumping. *Opt. Express* **2016**, *24*, 2215–2221. [CrossRef] [PubMed]

37. Tan, M.; Iqbal, M.A.; Nguyen, T.T.; Rosa, P.; Krzczanowicz, L.; Phillips, I.D.; Harper, P.; Forysiak, W. Raman Amplification Optimization in Short-Reach High Data Rate Coherent Transmission Systems. *Sensors* **2021**, *21*, 6521. [CrossRef] [PubMed]

38. Rosa, P.; Tan, M.; Le, S.T.; Phillips, I.D.; Ania-Castañón, J.; Sygletos, S.; Harper, P. Unrepeatered DP-QPSK Transmission Over 352.8 km SMF Using Random DFB Fiber Laser Amplification. *IEEE Photonics Technol. Lett.* **2015**, *27*, 1189–1192. [CrossRef]

39. Galdino, L.; Tan, M.; Alvarado, A.; Lavery, D.; Rosa, P.; Maher, R.; Ania-Castanon, J.D.; Harper, P.; Makovejs, S.; Thomesn, B.C.; et al. Amplification schemes and multi-channel DBP for unrepeatered transmission. *J. Light. Technolol.* **2016**, *34*, 2221–2227. [CrossRef]
40. Rosa, P.; Rizzelli, G.; Pang, X.; Ozolins, O.; Udalcovs, A.; Tan, M.; Jaworski, M.; Marciniak, M.; Sergeyev, S.; Schatz, R.; et al. Unrepeatered 240-km 64-QAM Transmission Using Distributed Raman Amplification over SMF Fiber. *Appl. Sci.* **2020**, *10*, 1433. [CrossRef]

 *sensors*

Article

Performance Analysis of a Vertical FSO Link with Energy Harvesting Strategy

Carmen Álvarez-Roa [1], María Álvarez-Roa [1], Francisco J. Martín-Vega [1], Miguel Castillo-Vázquez [1], Thiago Raddo [1,2] and Antonio Jurado-Navas [1,*]

[1] University Institute of Telecommunication Research (TELMA), University of Málaga, CEI Andalucía Tech., E-29071 Málaga, Spain; calvarez@ic.uma.es (C.Á.-R.); m.alvarez@ic.uma.es (M.Á.-R.); fjmvega@ic.uma.es (F.J.M.-V.); miguelc@ic.uma.es (M.C.-V.); thiago@ic.uma.es (T.R.)

[2] Engineering, Modeling & Applied Social Sciences Center, Federal University of ABC, Avenida dos Estados 5001, Santo André 09210-580, SP, Brazil

* Correspondence: navas@ic.uma.es

Abstract: In this paper we investigate the application of free space optical (FSO) communications, energy harvesting, and unmanned aerial vehicles (UAVs) as key technology enablers of a cost-efficient backhaul/fronthaul framework for 5G and beyond (5G+) networks. This novel approach is motivated by several facts. First, the UAVs, acting as relay nodes, represent an easy-to-deploy and adaptive network that can provide line-of-sight between the base stations and the gateways connected to the core network. Second, FSO communications offer high data rates between the UAVs and the network nodes, while avoiding any potential interference with the 5G radio access networks. Third, energy harvesting in the optical domain has the potential to extend the UAVs' battery life. Nevertheless, the presence of atmospheric turbulence, atmospheric attenuation, and pointing errors in the FSO links severely degrades their performance. For this reason an accurate yet tractable modelling framework is required to fully understand whether an UAV-FSO backhaul/fronthaul network with energy harvesting can be applied. To this end, we consider a composite channel attenuation model that includes the effect of turbulence fading, pointing errors, and atmospheric attenuation. Using this model, we derive analytical closed-form expressions of the average harvested energy as a function of the FSO link parameters. These expressions can be used to improve energy harvesting efficiency in FSO link design. We have applied our proposed expressions to evaluate the energy harvested in vertical FSO links for a variety of real scenarios under a modified on-off keying (OOK) scheme optimized for energy harvesting. From the simulations carried out in this paper, we demonstrate that significant values of harvested energy can be obtained. Such performance enhancement can complement the existing deployment charging stations.

Keywords: BER; vertical link; pointing errors; FSO; OOK; energy harvesting; atmospheric turbulence

Citation: Álvarez-Roa, C.; Álvarez-Roa, M.; Martín-Vega, F.J.; Castillo-Vázquez, M.; Raddo, T.; Jurado-Navas, A. Performance Analysis of a Vertical FSO Link with Energy Harvesting Strategy. *Sensors* **2022**, *22*, 5684. https://doi.org/10.3390/s22155684

Academic Editors: Yang Yue, Jiangbing Du, Jian Zhao and Yan-ge Liu

Received: 2 June 2022
Accepted: 26 July 2022
Published: 29 July 2022

1. Introduction

1.1. Overview of FSO Communications

The combination of the high bandwidth of optical communications with the flexibility of wireless technologies offered by free-space optical (FSO) communications has led to a fresh wave of innovation and research activities in this field [1]. Aside from the higher bandwidth, FSO communications offer several advantages over classical RF based wireless communications. First of all, the extremely high carrier frequencies inherent to the optical links make FSO detectors immune to multipath fading which severely degrades the performance of RF links [2,3]. Furthermore, FSO technology has the potential to reduce cost and consumed energy. The spectrum above 300 GHz is unlicensed, so operators do not have to pay for exclusive access in optical bands. Moreover, the components used in FSO links are cheaper, smaller, lighter and have lower power consumption as compared to that of

RF components [4]. More importantly, FSO technology does not interfere with RF systems, which paves the way to mixed RF/FSO approaches where the optical links complement the existing RF infrastructure [5,6]. Due to all these benefits, FSO communications have found their place in diverse applications such as space communications [7–9], underwater communications [10–13], indoor local area networks [14], data center networks [15], and mobile backhaul [6,16,17].

However, wireless optical links are affected by many impairments that may compromise their performance, so an accurate channel modeling is needed to anticipate the potential benefits of FSO-based approaches. These impairments are mainly categorized into four different effects: namely, pointing errors (i.e., misalignment losses), atmospheric losses, atmospheric turbulence (also called scintillation), and noise [1].

Thus, pointing errors are related to the misalignment of the transmit beam with respect to the field of view of the receiver. For fixed links of a few hundred meters, increasing the beam divergence can alleviate the misalignment loss at the expense of a higher geometric loss. Nevertheless, longer link distances or links with moving nodes require appropriate pointing, acquisition, and tracking procedures to mitigate the adverse effect caused by pointing errors [18].

Next, atmospheric loss is due to the presence of particles that either absorb or scatter the transmitted light (in that latter case, only the fraction of light scattered out of the location of the receiver is considered). In particular, the particles that affect FSO communication systems include those ones associated with rain, snow, fog, pollution, and smoke, among others.

On the other hand, the atmospheric turbulence or scintillation, is explained by inhomogeneities in the pressure and the temperature of the atmosphere that induce variations of the air refractive index along the transmission path [19]. These fluctuations cause random variations in the amplitude and phase of the received signal, i.e., fading, that lead to a considerable degradation in long-distance links. Several statistical models have been proposed in the literature to fully characterize the turbulence fading. The log-normal [20] distribution has been accepted as an accurate model for weak turbulence conditions whereas negative exponential [21] and Rayleigh [22] distributions have been used to model strong turbulence. For this reason, there have been remarkable research efforts to establish a common statistical model to characterize any turbulence condition. In this context the Gamma-Gamma distribution represents an appealing model that allows tractable analysis of the link performance while modelling turbulence conditions ranging from weak to strong turbulence [21].

Finally, the last adverse effect in FSO links is produced by noise. Background noise is the dominant one in most optical links. This noise is present due to the fact that the receiver collects some undesirable radiations such as reflected or scattered sunlight from hydrometeors or other objects. This radiation is mitigated by means of narrow spectral bandpass and spatial filtering prior to the photo-detector (PD). However, a non-negligible background noise might fall within the spatial and frequency ranges of the detector causing a random electrical signal that is added to the desired signal. This noise term can be modeled according to a Poisson distribution [23]; nevertheless, when the number of received photons associated with this background radiation is high enough, the Poisson distribution can be approximated by a Gaussian distribution [24]. These facts motivate the inclusion of an additive white Gaussian noise (AWGN) model in FSO links.

1.2. Related Work

The advances in pointing, acquisition, and tracking [18] have enabled the application of FSO communication to unmanned aerial vehicles (UAVs), which can act as relay nodes in 5G and beyond (5G+) cellular networks. This approach is especially promising in the context of backhaul/fronthaul networking [25] since the mobility of UAVs and the height at which they operate provides a reliable line-of-sight link for a high-bandwidth FSO connection [6]. In addition, the location of the relay nodes can be changed, making this kind of network adaptable to changes in weather and traffic needs. For instance, if any base station (BS) is highly loaded, UAVs can connect to that BS to readjust the backhaul traffic. Accordingly, if the atmospheric loss conditions worsen, e.g., due to fog, UAVs could approach the BSs to maintain a reliable FSO data connection.

However, one of the main drawbacks associated with extending the network with aerial access points is related to the service interruption when UAVs land to recharge their batteries. UAVs need to fly back to a nearby charging station to recharge their on-board battery frequently. For this reason, energy-efficient frameworks are preferred to extend the service time provided by UAVs. To this end, two different solutions (or a combination of both) are normally proposed: (i) trajectory optimization; and (ii) energy harvesting. The former approach consists on a route design to minimize the consumed energy while guaranteeing a target rate with a given node [26,27]. On another note, energy harvesting (EH) involves capturing, and storing energy from external sources, e.g., solar power, thermal energy, wind energy, or kinetic energy, which is generally known as ambient energy [28]. Interestingly, we can remark on the simultaneous lightwave information and power transfer (SLIPT), which is a kind of EH, where the captured energy comes from the optical signal that carries the information [29]. This latter approach is very promising since the optical signal uses narrower beams, and thus the emitted energy is concentrated towards the receiver, which makes SLIPT systems particularly efficient. Despite their relevant and potential benefits to extend the UAVs battery life, the number of works focused on the application of SLIPT for FSO-based UAVs nodes is limited. For the sake of clarity, some of these works are described below.

To start with, in [30] a two hop, mixed FSO-RF relaying scheme is proposed and analyzed assuming Gamma-Gamma turbulence fading and pointing errors. Under these assumptions, the outage probability in terms of the bivariate Fox-H function is derived. The proposed scheme considers a first hop in the optical domain where the relay can capture the energy from the received light. Then, this energy is used for RF signal transmission. Results reveal that a larger PD responsivity results in a better performance.

A pioneering work about the application of FSO with EH to UAVs-based networks for 5G backhauling is detailed in [6], and assessed via simulations in terms of capacity and cost. Simulation results are obtained for various different conditions involving atmospheric loss, turbulence loss, and pointing errors. Further investigation was performed in [31] where the SLIPT scheme was posited as an optimization problem to maximize the harvested energy while guaranteeing a target rate, for different typical modulations such as pulse amplitude modulation (PAM) and pulse position modulation (PPM).

A novel view about the SLIPT communication between BSs and UAVs nodes was presented in [32]. This work relies on mathematical tools from stochastic geometry to analyze the performance of UAVs based networks. With this approach, the main metrics are obtained as spatial averages over infinite realizations involving different BSs and UAVs locations. To assess the effectiveness of EH and FSO based communication, the joint energy and the signal-to-noise ratio (SNR) coverage probability are derived, i.e., the joint probability of the UAV receiving enough energy to ensure successful operation (hovering and communication) and having a received SNR higher than a given threshold.

To finish this brief review, we must cite the work presented in [33] where a mixed FSO/RF network with EH was proposed. In this scheme, ground stations transmit backhaul traffic to the UAVs, which act as moving BSs. Subsequently, the UAVs transmit RF signals to the ground users. Accordingly, the UAVs harvest and store the received energy coming from the ground stations through their FSO links. The proposed framework is considered as an optimization problem to maximize the energy efficiency.

1.3. Contributions

Despite their relevance, none of the aforementioned studies, e.g., [6,29–33], consider two important aspects: (i) the random nature of the EH, which is inherited from the variability of the optical channel; and (ii) the trade-off between bit error rate (BER) and EH.

In this paper, we investigate the influence of this random behavior on the design of EH-optimized links, while fulfilling a target BER. This will be achieved by transmitting a certain fraction, ζ, of the peak transmission power, P_m, with the logical symbol '0'. As shown in this paper, a proper design of the ζ maximizes the harvested energy while fulfilling a desirable reliability in terms of the BER.

More specifically, we aim to respond to the following inquiries: (i) the optimum ζ value to maximize the harvested energy while fulfilling a target BER; and (ii) the maximum average harvested energy that can be achieved under realistic link conditions. To address those two issues we derive some closed-form expressions for the two main metrics here considered: the average BER and the average harvested energy. Therefore, the paper's contributions can be summarized as follows:

1. We propose an accurate mathematical model for the UAV-based approach to provide backhaul connectivity for 5G+ networks [6]. To this end, ground stations and UAVs establish FSO links with EH to recharge the UAVs batteries and extend the service time. Furthermore, the UAVs will have another FSO link with the BSs to provide backhaul connectivity. The mathematical model considers realistic channel impairments such as turbulence fading (scintillation), atmospheric loss, pointing errors, and additive white Gaussian noise (AWGN).
2. We derive analytical closed-form expressions of the average harvested energy as a function of the FSO link parameters. These expressions can be used to improve energy harvesting efficiency of the FSO links.
3. Closed-form expressions for the BER have been obtained for links using a variant of the on-off Keying (OOK) scheme that has been properly modified to maximize the EH. With this scheme, which is referred to as OOK-EH, a certain fraction, ζ, of the peak optical power, P_m, is also transmitted with the symbol '0' to recharge the batteries of the UAV.
4. We obtain the optimal ζ value that maximizes the average harvested energy while maintaining its associated BER smaller than a *target* BER. Such a target BER is defined in practical communication standards to guarantee a reliable transmission.
5. The performance associated with the proposed scheme is corroborated with extensive Monte Carlo (MC) simulations in diverse and realistic atmospheric conditions. An excellent agreement between theoretical and simulation results is observed. Results reveal that the combination of FSO, EH, and UAVs provide reliable backhaul links while capturing energy to extend the UAVs service time. Let us recall that short lifetime of the batteries mounted on drones is seen as the main limitation of the proposed approach. For this reason, one of the main targets of our work is to obtain the optimal amount of energy that can be collected to charge the batteries of the UAVs involved in the system, as discussed in [30,31], while maintaining its performance in terms of BER.

The rest of the paper is structured as follows. In Section 2 the system model is mathematically described, including the signal as the channel model. Then, the main metrics are presented and derived in Section 3, namely, the expressions for the amount of harvested energy by the system and its BER. In Section 4 numerical results are obtained to assess the benefits of the proposed scheme, which allows to identify both trends and limiting factors. Finally, relevant conclusions are discussed in Section 5.

2. System Model

2.1. Proposed Scenario

We consider a flying backhaul/fronthaul network composed of UAVs that provides reliable connectivity to a 5G+ radio access network (RAN). The UAVs act as networked flying platform (NFP) nodes that can react to changes in weather or traffic conditions [34]. Its ground-to-air links are based on FSO technology with EH. This solution has four remarkable benefits: (i) its FSO links provide the required high data rates for the backhaul or fronthaul connections; (ii) FSO links do not cause interference to the 5G RAN;

(iii) UAVs can adapt their location to the traffic and channel conditions; and (iv) EH extends the UAVs' service time and thus, improves the network performance.

This approach is illustrated with Figure 1 where it shows the representative nodes and connections. The backhaul traffic represents the IP data transmissions between the 5G core network and the BSs in distributed RAN approaches [35]. Accordingly, the BSs fully implement the 5G RAN protocol stack, and thus they are source/destination of IP packets towards/from the core network.

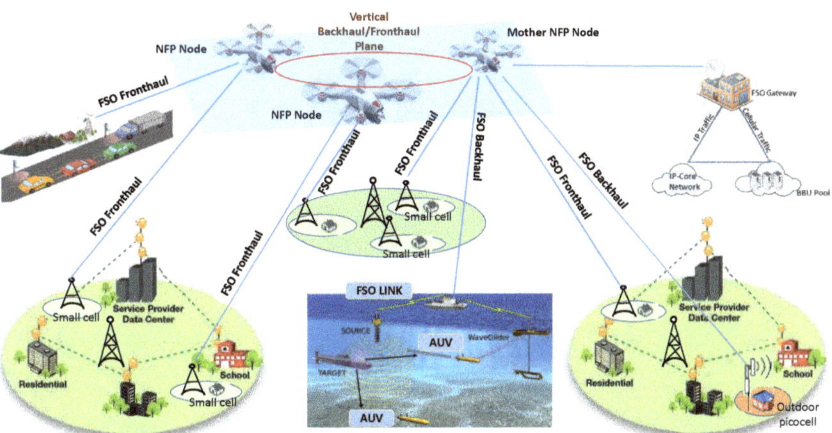

Figure 1. Possible 5G + network scenarios where FSO will extend or complement existing deployments.

A recent approach to the classical distributed RAN arises with the cloud RAN (C-RAN) [36], where a base band unit (BBU) centralizes the RAN processing of many small cells. The signal transmission related to each cell is carried out by radio remote heads (RRHs) that are connected with the BBUs through fronthaul links. This latter approach offers two main benefits compared to distributed RAN. First, C-RAN reduces costs since each small cell can be implemented with a RRH, cheaper than a complete BSs. Second, C-RANs ease the implementation of coordinated mechanisms for interference mitigation.

Coming back to Figure 1, it is straightforward to see the different backhaul links between the core network and the BSs through the FSO-UAVs based network. This scheme requires a FSO gateway that creates the link with a mother NFP node, i.e., a mother UAV that forwards the IP traffic towards the NFP node currently connected to the target BS. Analogously, the fronthaul links between the BBU and the RRHs also require a connection between a FSO gateway and the mother NFP node. Certainly, many applications beyond classical mobile networks can benefit from the proposed scenario shown in Figure 1 since it can provide backhaul/fronthaul connectivity to 5G V2X, or even future underwater communications [11–13,37].

2.2. Received Signal Model

Throughout this paper, we consider a non-coherent intensity modulation with direct-detection (IM/DD) scheme since it is a practical and widely extended technique for FSO communications [38]. With this scheme, the intensity of the emitted light is employed to convey the information. Therefore, the PD output current can be expressed as follows [39,40]

$$i = hRx + n, \tag{1}$$

where x is the transmitted intensity in W, R is the detector responsivity in W/A, h is the channel coefficient and n represents the noise term at the receiver side. We assume that the shot noise induced by the ambient light is the dominant noise source in our analysis. Thus, we model n as a signal-independent zero-mean Gaussian noise with variance given by

$$\sigma_n^2 = 2qRB_n P_{amb},\tag{2}$$

where q denotes the electron charge, B_n is the receiver noise bandwidth, and P_{amb} is the ambient light power. This latter term can be obtained as $P_{amb} = \pi r^2 S_n B_o \Omega_{FOV}$, where r is the receiver aperture radius, S_n is the ambient light spectral radiance, B_{opt} is the optical bandwidth in nm and Ω_{FOV} is the receiver field of view (FOV) in srad.

The transmitted intensity is taken as symbols drawn with equal probability from the OOK constellation such that $x \in \{P_1, P_0\}$, where P_1 and P_0 represent the power corresponding to the transmission of a '1' and a '0', respectively.

In this respect, we can cite three primary atmospheric phenomena that affect optical wave propagation and that constitute the total channel coefficient h: (i) the deterministic path loss, h_l characterized by absorption and scattering; (ii) the geometric spread and pointing errors h_p; and (iii) the refractive-index fluctuations (i.e., the atmospheric turbulence), h_a, leading to irradiance fluctuations. Thus, the total channel attenuation is modeled as the product of these aforementioned channel factors as:

$$h = h_l h_p h_a,\tag{3}$$

where h_p and h_a are random variables (RVs). In the next sections, we describe the underlying distribution that model those RVs. Finally, we define the signal to noise ratio as [41]:

$$\text{SNR} = \frac{(RhP_{av})^2}{\sigma_n^2},\tag{4}$$

where $P_{av} = (P_0 + P_1)/2$ represents the average transmit power.

2.3. Atmospheric Turbulence Model

In order to model the intensity fluctuations caused by the atmospheric turbulence, the statistical Gamma-Gamma distribution is here assumed because of its mathematical tractability and accuracy to characterize a wide variety of scenarios ranging from weak to strong turbulence. Thus, and following [21], the probability density function (pdf) of h_a is written as

$$f_{h_a}(h_a) = \frac{2(\alpha\beta)^{(\alpha+\beta)/2}}{\Gamma(\alpha)\Gamma(\beta)} h_a^{(\alpha+\beta)/2-1} K_{\alpha-\beta}\left(2(\alpha\beta h_a)^{1/2}\right),\tag{5}$$

where $K_p(x)$ is the modified Bessel function of the second kind, and $\Gamma(x)$ is the Gamma function, with α representing the effective number of large-scale cells of the scattering process and with β denoting the effective number of small-scale cells. Namely, from [39], these latter parameters, α and β, can be obtained as

$$\alpha = \left[\exp\left(0.49\sigma_R^2(1 + 1.11\sigma_R^{12/5})^{-7/6}\right) - 1\right]^{-1},\tag{6}$$

and

$$\beta = \left[\exp\left(0.51\sigma_R^2(1 + 0.69\sigma_R^{12/5})^{-5/6}\right) - 1\right]^{-1},\tag{7}$$

respectively, where σ_R^2 is the Rytov variance which, for uplinks paths, is defined as [39]

$$\sigma_R^2 = 2.25k^{7/6}(Z - z_0)^{5/6}\int_{z_0}^{Z} C_n^2(z)\left(\frac{z - z_0}{Z - z_0}\right)^{5/6} dz,\tag{8}$$

with $C_n^2(z)$ being the index of refraction structure parameter at altitude z, whereas $k = 2\pi/\lambda$ is the optical wave number and Z and z_0 are the UAV and transmitter heights, respectively.

2.4. Atmospheric Attenuation

For optical waves, the effect of the atmospheric attenuation suffered by the light propagating through the atmosphere is mainly caused by either absorption as scattering by air molecules in addition to both absorption and scattering by solid or liquid particles suspended in the air which, as a last resort, indicates the effect of weather conditions on the transmitted laser beam. Absorption and scattering are often grouped together under the topic of extinction, defined as the reduction or attenuation in the amount of radiation passing through the atmosphere. Mathematically speaking, such attenuation is incorporated using the well-known Beer–Lambert law [42–44] given by:

$$h_l = \exp\left(-a(\lambda)L\right), \tag{9}$$

where $a(\lambda)$ is the wavelength dependent attenuation coefficient (extinction coefficient) and $L = Z - z_0$ is the propagation path length from the transmitter. As commented, the coefficient a depends on weather conditions and can be obtained from the atmospheric visibility. Since the attenuation coefficient hardly changes over long periods of time, we have assumed h_l as a deterministic coefficient in our analysis.

2.5. Geometric Spread and Pointing Error Model

In addition to attenuation and atmospheric turbulence, geometric beam spread and pointing accuracy also affect the performance of these systems. The geometric loss is caused by the divergence of the transmit beam when propagating through the atmosphere, as $\omega_z = \theta_T \cdot L$, with ω_z being the received beam waist, with θ_T denoting the transmitter divergence angle, whereas L is the propagation path length. Since the received beam width is usually wider than the lens aperture size, part of the transmitted power cannot be collected, leading to loss. Thus, this geometric loss depends mainly on the ratio between the received beam waist, ω_z, and the receiver aperture radius, r. However, during the designing of a FSO link, it is possible to control the beamwidth produced at a certain distance, and therefore the ratio ω_z/r, by adjusting properly the laser parameters.

On the other hand, imperfections in the pointing, acquisition, and tracking process between the ground stations and the UAVs can also cause loss. Note that both phenomena are interrelated and, thus, accurate modeling frameworks have been proposed to account for both effects. To this end, in this work, we consider the general model proposed in [44,45]. According to this model, a Gaussian beam profile is assumed with a beam waist, ω_z, on the receiver plane and a circular aperture receiver of radius r; then, the attenuation due to the geometric spread with pointing error can be approximated as the Gaussian form

$$h_p \approx A_o \exp\left(-\frac{2\rho^2}{\omega_{zeq}^2}\right), \tag{10}$$

where ρ, is the radial pointing error, A_o is the fraction of collected power without pointing error, i.e., only due to geometric spread, and ω_{zeq}^2 is the equivalent beam width. Here, A_o and ω_{zeq}^2 are given by $A_o = \mathrm{erf}(\mu)^2$ and $\omega_{zeq}^2 = \omega_z^2 \sqrt{\pi}\,\mathrm{erf}(\mu)/[2\mu \exp(-\mu^2)]$, respectively, being $\mathrm{erf}(\cdot)$ the error function and $\mu = \sqrt{\pi}r/(\sqrt{2}\omega_z)$. Moreover, considering independent identical Gaussian distribution for the horizontal x and y displacement in the receiver plane, the radial error $\rho = \sqrt{x^2 + y^2}$ is modeled as Rayleigh distribution with a jitter variance at the receiver σ_s^2. Under these assumptions, the channel attenuation, h_p, can be seen as a function of the radial displacement, ρ, which is a RV. Hence, the pdf of h_p can be seen as a random variable transformation problem, which leads to the following expression [45]:

$$f_{h_p}(h_p) = \frac{\gamma^2}{A_o^{\gamma^2}} h_p^{\gamma^2 - 1}, \tag{11}$$

where $\gamma = \omega_{zeq}/(2\sigma_s)$ denotes the ratio between the equivalent beam radius at the receiver and the pointing error displacement standard deviation.

2.6. Composite Channel Model

Once the three factors included in Equation (3) have been individually discussed, the statistical characterization of the composite channel $h = h_a h_l h_p$ can be achieved. Thus, from [45],

$$f_h(h) = \frac{2\gamma^2(\alpha\beta)^{\frac{\alpha+\beta}{2}}}{(A_o h_l)^{\gamma^2}\Gamma(\alpha)\Gamma(\beta)} h^{\gamma^2-1} \int_{\frac{h}{A_o h_l}}^{\infty} h_a^{\frac{\alpha+\beta}{2}-1-\gamma^2} K_{\alpha-\beta}\left(2\sqrt{\alpha\beta h_a}\right) dh_a. \tag{12}$$

where α and β parameters include the information on the strength of the turbulence, with γ containing the severity of the pointing error, where A_o denotes the geometric spread attenuation and with h_l being the path loss.

3. System Performance

In this paper, two metrics are considered to evaluate the performance of FSO vertical links. First, the amount of harvested energy in the UAV is analyzed through the average EH parameter. Second, the received signal quality is studied in terms of the bit error rate (BER) parameter.

In particular, we must remark that the harvested energy analysis detailed below is valid for any modulation scheme; nevertheless, our BER analysis will be particularized for a variant of the OOK scheme previously optimized for energy harvesting under a fixed peak power constraint [31]. An OOK scheme was selected because of its simplicity and low power consumption, especially interesting due to limited on-board UAVs battery lifetime. In such a scheme, named OOK-EH, a power $P_1 = P_m$ is emitted for the transmission of a logical '1', where P_m is the peak power of the laser; whilst a fraction of P_m is carried for transmitting a logical '0', i.e.,

$$P_0 = \zeta P_m, \tag{13}$$

where $0 \leq \zeta < 1$. In contrast to the classical OOK, where no light is emitted when a logical '0' is transmitted, our proposed scheme always carries a power level that will depend on the ζ parameter. This added DC component increases the average optical power P_{av} and can be used to improve energy collection by adjusting the ζ parameter appropriately. Furthermore, and in order to maximize the harvested energy, a power level P_m is also emitted when the transmitter is idle. However, any increase in the P_0 level will reduce the dynamic range given by the peak average optical power ratio (PAOPR) and, consequently, its associated BER will be increased. The PAOPR is given by $2/(1+\zeta)$. Consequently, there is an important trade-off between increasing harvested energy and degrading BER. For any particular set of channel conditions, this trade-off is achieved for an optimal ζ that maximizes the harvested energy while keeping the BER performance below a predetermined target BER.

In order to evaluate the performance of any vertical FSO link with our proposed OOK-EH technique under realistic channel conditions, analytical closed-form expressions for the average EH and BER are derived in this section, considering the aforementioned main phenomena degrading the quality of the received signal, i.e., noise, turbulence, path loss, geometric spread and pointing errors.

3.1. EH Performance

An EH module is added to the receiver to collect the energy from the received signal. This module extracts the DC component, I_{DC}, of the electrical signal that is obtained by the PD, i. The DC current is then either stored or directly used to feed other modules such as the information detection module and the UAV's motor, which translates into extending the UAV's battery lifetime. Figure 2 illustrates a block diagram of the receiver where the EH and information detection modules are shown. As it can be observed, the EH module is formed by a Schottky diode and a low-pass filter (LPF), which are passive devices, and thus, its associated power consumption is limited and could be included in the conversion efficiency, ζ [46]. Furthermore, this approach does not require either any control logic or any additional module to obtain the DC current.

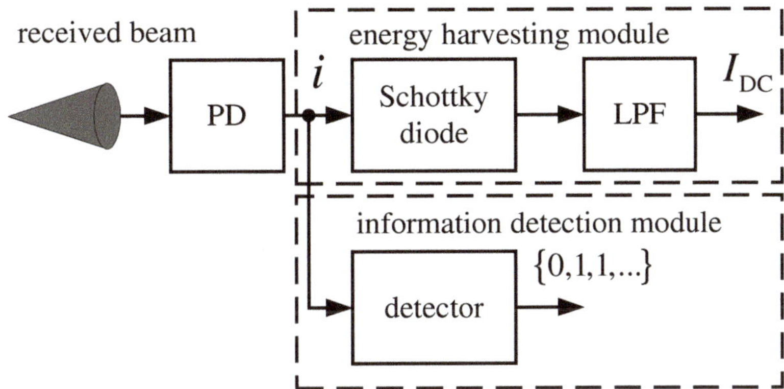

Figure 2. Block diagram of the proposed receiver structure with EH. An optical circular converging lens would be placed in front of the PD.

As described in [29,31,47], the harvested energy per second in the optical receiver is given by

$$EH = fV_t I_{DC} \ln\left(1 + \frac{I_{DC}}{I_o}\right), \tag{14}$$

where f, V_t and I_o stand for the photo-detector's fill factor, thermal voltage, and dark saturation current, respectively. The DC component of the output current, I_{DC}, is written as [47]:

$$I_{DC} = RhP_{av}. \tag{15}$$

Here, R denotes the PD responsivity in A/W, with h being the composite channel attenuation coefficient, whereas P_{av} represents the average transmitted power.

As can be noticed, the harvested energy is random due to the influence of the channel attenuation coefficient h. Therefore, to obtain the collected energy over a long period of time, the average EH (AEH) is derived as:

$$AEH = \int_{h>0} f_h(h) f V_t RhP_{av} \ln\left(1 + \frac{RhP_{av}}{I_o}\right) dh, \tag{16}$$

where $f_h(h)$ is the pdf of the composite channel attenuation according to (12). Employing the expressions [48] (Eqs. (07.34.21.0011.01) and (07.34.21.0085.01)), the solution of this integral can be written in a closed-form as

$$AEH = \frac{\gamma^2 f V_t RP_{av} A_o h_l}{\alpha\beta\Gamma(\alpha)\Gamma(\beta)} G_{5,3}^{1,5}\left(\frac{A_o h_l RP_{av}}{I_o \alpha\beta} \middle| \begin{array}{c} 1,1,-\gamma^2,-\alpha,-\beta \\ 1,-1-\gamma^2,0 \end{array}\right). \tag{17}$$

Moreover, when only the effect of turbulent fading is considered, and $f_h(h)$ is given by (5), the above equation reduces to

$$AEH = \frac{f V_t RP_{av}(A_o h_l)^2}{(\alpha\beta)^2 \Gamma(\alpha)\Gamma(\beta)} G_{5,3}^{1,5}\left(\frac{A_o h_l RP_{av}}{I_o \alpha\beta} \middle| \begin{array}{c} 1,1,-1,-1-\alpha,-1-\beta \\ 1,-2,0 \end{array}\right). \tag{18}$$

where the term P_{av}, which is present in Equations (16)–(18), is expressed as $P_{av} = P_m(1+\zeta)/2$ in case of OOK-EH. From Equations (17) and (18), the amount of harvested energy is not depending on the modulation scheme itself but on the average transmitted power. Of course, the particular channel conditions and the type of PD and rest of modules shown in Figure 2 will influence on the amount of energy any UAV can extract.

3.2. EH Optimization

In view of the expressions for the average harvested energy derived above, it is straightforward to observe that they depend on: first, the channel and link conditions (i.e., attenuation and turbulence, fundamentally); second, on the transmitter and receiver systems with a given pointing misalignment loss, geometric spread, and maximum peak power, P_m; and third, on the parameter ζ. In this section, we consider all these parameters as uncontrollable variables that depend on the channel and the physical devices, except for the ζ variable, that can be adjusted to maximize the harvested energy.

Therefore, we notice that if ζ increases, then the average harvested energy also increases; however, this fact adversely affects on the BER performance of the system. In this respect, there exists an interesting trade-off between EH and reliability. We consider in this work that the communication system includes a simple forward error correction (FEC) decoder, which defines a maximum pre-FEC BER, normally referred to as $\text{BER}_{\text{target}}$. Such a target is the maximum BER of the coded bits before the FEC decoder to guarantee an error-free transmission with high probability.

We assume two widely adopted target BER values of $\text{BER}_{\text{target}} = 5 \times 10^{-5}$ and $\text{BER}_{\text{target}} = 10^{-8}$ related to different FEC options as defined in [49]. Interestingly, the latter target BER is related to a BASE-R code, which is also know as Fire Code FEC (FC-FEC) as is defined in (clause 74 of [49]). This latter code requires a very small computational complexity for its decoding.

Thus, we pose the optimization problem for the design of the proposed FSO system with EH as follows

$$\arg\min_{\zeta \in [0,1)} \text{AEH}$$
$$\text{s.t. BER} < \text{BER}_{\text{target}} \tag{19}$$

The above problem is solved numerically in Section 4 to determine the optimal ζ value and assess the maximum AEH that can be obtained with the proposed framework under realistic and diverse channel conditions.

3.3. BER Performance

Now, we derive closed-form expressions of the BER for the OOK scheme used for SLIPT. Hence, the received BER can be written as

$$\text{BER} = p_0 P_e(1|0) + p_1 P_e(0|1), \tag{20}$$

where $P_e(1|0)$ and $P_e(0|1)$ represent the probability of false alarm and the probability of missed detection, respectively. In our analysis, we assume equally-likely symbols, $p_0 = p_1 = 0.5$ in many different realistic scenarios.

First, for an ideal scenario without neither turbulence nor pointing error (in (3), h is reduced to $h = A_o h_l$), the terms $P_e(1|0)$ and $P_e(0|1)$ can be expressed as [50]:

$$P_e(1|0) = \frac{1}{2}\mathrm{erfc}\left(\frac{i_t - i_{s_0}}{\sqrt{2\sigma_n^2}}\right) = \frac{1}{2}\mathrm{erfc}\left(\frac{i_t - i_{s_1}\zeta}{\sqrt{2\sigma_n^2}}\right) \tag{21}$$

$$P_e(0|1) = \frac{1}{2}\mathrm{erfc}\left(\frac{i_{s_1} - i_t}{\sqrt{2\sigma_n^2}}\right); \tag{22}$$

with i_{s0} and i_{s1} denoting the received photocurrent corresponding to the transmission of a logical '0' and a logical '1', respectively; where i_t is the optimal detection threshold and σ_n^2 is the noise variance given in Equation (2). On another note, and from (13), the photocurrents i_{s0} and i_{s1} can be expressed as $i_{s0} = Rh\zeta P_m$ and $i_{s1} = RhP_m$, respectively. In addition, i_t can be written as a function of ζ as

$$i_t = \frac{i_{s_1} + i_{s_0}}{2} = \frac{i_{s_1}(1 + \zeta)}{2}. \tag{23}$$

Recall that, for this ideal AWGN channel, h has a deterministic behavior since neither turbulence-induced nor misalignment fadings were considered yet. Accordingly, its associated BER is obtained by substituting (21) and (22) into (20), and it is given by

$$\mathrm{BER} = \mathrm{erfc}\left(\frac{RhP_m(\zeta - 1)/2}{\sqrt{2\sigma_n^2}}\right). \tag{24}$$

Second, we consider a more realistic scenario where atmospheric turbulence is now considered, although misalignment fading is still not included Therefore, the channel attenuation coefficient, h, becomes a RV since the turbulence-induced scintillation now incorporated into h follows the Gamma-Gamma pdf shown in (5). Thus (24) must be averaged with the corresponding pdf describing the behavior of h, and [48] (Eqs. (07.34.21.0013.01) and (07.34.21.0085.01)) are, again, used to solve the resulting integral:

$$\mathrm{BER} = \frac{2^{(\alpha+\beta)}}{8\pi\sqrt{\pi}\Gamma(\alpha)\Gamma(\beta)} G_{5,2}^{2,4}\left(\frac{8(RA_o h_l P_m(\zeta - 1)/2)^2}{(\sigma_n\alpha\beta)^2}\,\middle|\,\begin{matrix} \frac{1-\alpha}{2}, \frac{2-\alpha}{2}, \frac{1-\beta}{2}, \frac{2-\beta}{2}, 1 \\ 0, \frac{1}{2} \end{matrix}\right) \tag{25}$$

Finally, when pointing errors are also taken into account, (21) and (22) must be now averaged with respect to (12). Hence,

$$\mathrm{BER} = \frac{2^{(\alpha+\beta)}\gamma^2}{16\pi\sqrt{\pi}\Gamma(\alpha)\Gamma(\beta)}$$
$$G_{7,4}^{2,6}\left(\frac{8(RA_o h_l P_m(\zeta - 1)/2)^2}{(\sigma_n\alpha\beta)^2}\,\middle|\,\begin{matrix} \frac{1-\gamma^2}{2}, \frac{2-\gamma^2}{2}, \frac{1-\alpha}{2}, \frac{2-\alpha}{2}, \frac{1-\beta}{2}, \frac{2-\beta}{2}, 1 \\ 0, \frac{1}{2}, \frac{-\gamma^2}{2}, \frac{1-\gamma^2}{2} \end{matrix}\right), \tag{26}$$

after having used [48] (Eqs. (07.34.21.0013.01) and (07.34.21.0085.01)). As a previous step, either the complementary error function, ($\mathrm{erfc}(\cdot)$), and the Bessel's K function were expressed in terms of Meijer's G functions from [48] (Eqs. (03.04.26.0006.01) y (06.27.26.0006.01)).

As can be seen from these latter equations, the exact expression for the error probability is given in terms of Meijer's G-function, which may be difficult to facilitate further analytical studies. Hence, it would be possible to obtain simpler expressions after some mathematical approximations following the approximation given in [51] involving an upper bound and a lower bound for the Gaussian Q-function that represents the behavior in terms of error probability of an ideal Gaussian channel, as shown in Equation (22), and based on series expansion. Additionally, and for a future work, we have planned to use a more generic model for the turbulence, the Málaga model [52] and its formulation from Generalized-K functions [53]. Thus, as commented in [54], its pdf can be approximated by a Gauss–Laguerre quadrature. Since the Gamma-Gamma model employed in our paper is a particular case of the Málaga model [52], then this Gauss–Laguerre quadrature can also be applied. Thus the upper bound given in [54] and [51] (Equation (19)) can be employed.

4. Results and Discussions

In this section we analyze the impact of FSO link parameters on system performance in terms of energy harvesting and quality using the AEH and BER expressions derived in the previous section. This analysis allows to identify which are the key design parameters and the trade-offs to be considered in order to properly choose such parameters.

For our numerical analysis, we have assumed the values given in Table 1. In particular, we consider a vertical FSO link consisting of a ground-based optical transmitter located at a height z_0 and a receiver mounted on a UAV hovering on the ground at a height Z. Therefore, the separation between both is $L = Z - z_0$. The transmitter transmits a light beam with a divergence angle θ_T = 1 mrad [6] and a peak power P_m operating at a wavelength of 1550 nm. The receiver, as in [44], is composed of an optical circular converging lens whose effective light collection area is characterized by an aperture radius, r, of 10 cm with a responsivity $R = 0.5$ A/W. Some commercial implementations following these features can be found at [55,56] Therefore, for $L = 300$ m [57], the ratio of the beam waist to the aperture radius of the receiver is $(\omega_z/r) = 3$ (remember that a source with a 1 mrad divergence angle was considered) while, for $L = 1000$ m, $(\omega_z/r) = 10$. To cover a large area on the ground, a field of view FOV$= 45°$ is assumed. Due to this large FOV, shot noise caused by ambient light is the dominant source of noise in the receiver. A spectral radiance $S_n = 1$ mW/(cm^2 nm srad), with an optical bandwidth $B_o = 10$ nm and a noise bandwidth $B_n = 1$ GHz are assumed to obtain σ_n^2 from (2).

Regarding the turbulence, we calculate the magnitudes of α and β, with the expressions (6) and (7), considering the values of $C_n^2(z)$ provided by the Hufnagel–Valley model [39] for different link heights, assuming $C_n^2(z_0) = 1.7 \times 10^{-13}$ m$^{-2/3}$. Thus, α and β vary between 25 and 30 for heights between 300 m and 1000 m, leading to $\sigma_R^2 << 1$, which corresponds to a very weak turbulence condition. In addition, we consider a jitter of $\sigma_s = 10$ cm [58] to model the pointing error.

Finally, we use the attenuation coefficients corresponding to very clear air, clear air, and haze shown in Table 1.

Table 1. System parameters.

FSO Parameters		
Parameter	**Symbol**	**Value**
Operating Wavelength	λ	1550 nm
Transmitter divergence	θ_T	1 mrad
Responsivity	R	0.5 A/W
Receiver aperture radius	r	10 cm
Field of view	FOV	45°
Optical Bandwidth	B_o	10 nm
Noise Bandwidth	B_n	1 GHz
Spectral radiance	S_n	1 mW/cm^2 nm srad
Turbulence, pointing error and climatic parameters		
Parameter	**Symbol**	**Value**
Structure parameter	$C_n^2(z_0)$	1.7×10^{-13} m$^{-2/3}$
Number of large-scale cells	α	30
Number of small-scale cells	β	30
Jitter variance	σ_s	10 cm
Attenuation coefficient (very clear air)	a	0.0647 dB/km
Attenuation coefficient (clear air)	a	0.2208 dB/km
Attenuation coefficient (haze)	a	0.7360 dB/km
EH parameters		
Parameter	**Symbol**	**Value**
Fill factor	f	0.75
Dark saturation current	I_o	10^{-9} A
Thermal voltage	V_t	25 mV

Figure 3 shows the typical attenuation of vertical optical links (Figure 3a) together with the potential energies harvested (Figure 3b) for different separation distances between any UAV and the ground BS, ranging between 200 m and 1000 m.

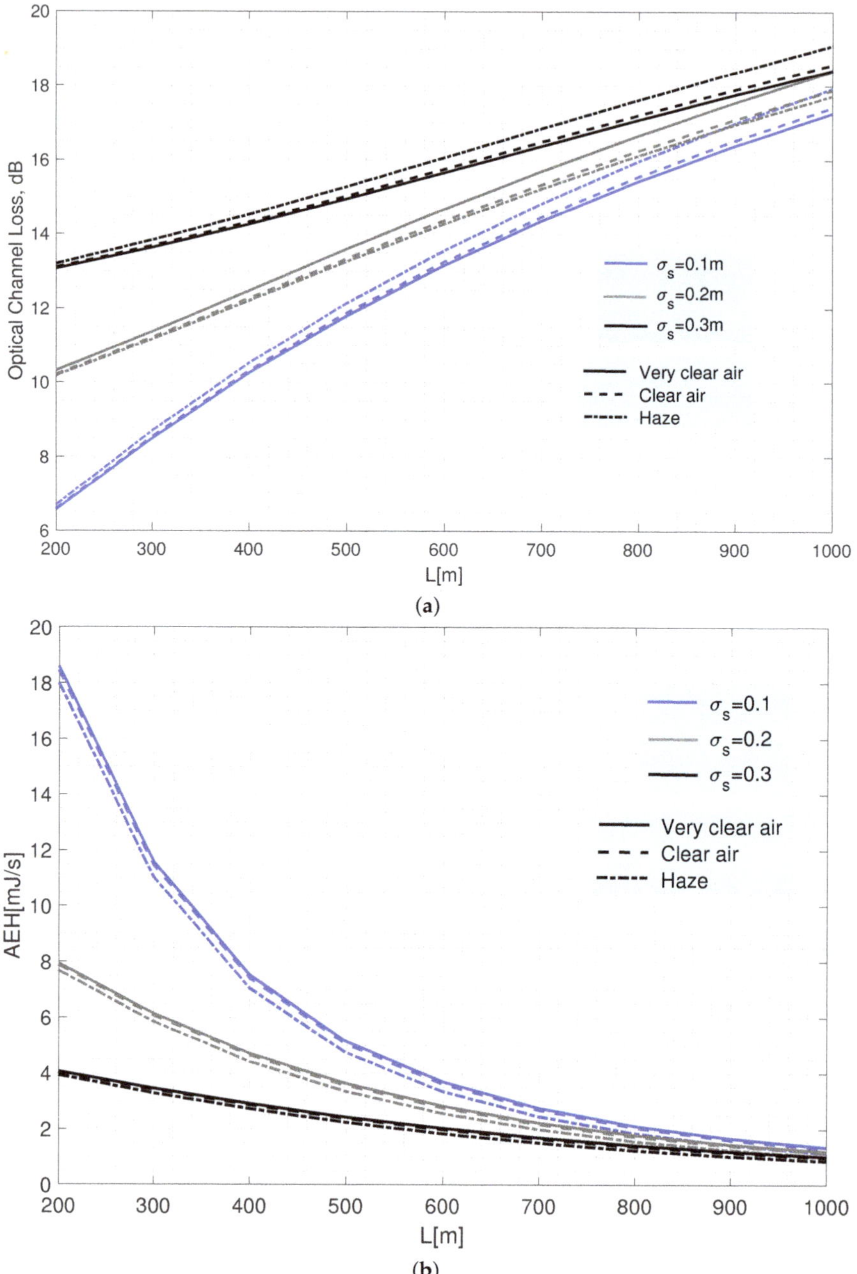

Figure 3. (**a**) Average optical channel loss as a function of link length. (**b**) Average harvested energy as a function of the link length. Both figures consider different UAV jitters and weather conditions and a beam divergence of 1 mrad. To obtain the AEH, a single transmitter with $P_{av} =500$ mW is assumed.

Channel loss depicted in Figure 3a is calculated by $-10\log(\bar{h})$, with $\bar{h} = (h_l A_o \gamma^2)/(1+\gamma^2)$ representing the average attenuation coefficient calculated from Equation (12). It is straightforward to check how the coverage provided by any deployed UAV enhances with increasing its altitude but, as shown in Figure 3a, at the cost of a significant increase in power losses at the receiver. Specifically, that increase is mainly caused by the beam broadening induced by the transmitter divergence. Namely, for $L =$300 m (i.e., $(\omega_z/r) =$3, as commented above), a power loss of 8.5 dB is reached for $\sigma_s = 0.1$ m. That value grows to 18 dB for $L =$1000 m as $(\omega_z/r) =$10. In both cases, a transmission divergence of 1 mrad and a 10 cm aperture radius were considered.

In addition, Figure 3b shows that increasing the channel loss dramatically reduces the harvested energy. Thus, for a single transmitter with a $P_{av} = 500$ mW, the AEH drops from 18 mJ/s for the best case at 200 m; to less than 2 mJ/s at 1000 m. Note that the AEH in the figure has been calculated using Equation (17). Finally, Figure 3 shows how atmospheric losses are not relevant for the amount of energy the UAV can harvest due to the short link lengths considered in the analysis. Furthermore, they can be neglected without loss of accuracy when the UAV is affected by UAV's jitter caused by its motor vibration.

Moreover, both figures also show the impact on the link performance of the pointing deviation caused by the jitter inherent to the UAV that acts as receiving side. For example, it can be observed that an increase in the UAV's jitter (higher σ_s) is more detrimental when flying at lower altitudes than at higher altitudes. The reason for this is that the ω_z/r ratio decreases for lower altitudes due to the beam divergence angle and, thus, the random variations inherent to the UAV location cause a more serious channel loss than in higher altitudes, even causing that the communication link may even be interrupted if the UAV moves out of the transmitted beam footprint. Consequently, energy harvesting is less affected by jitter as the UAV is operating at a higher altitude and, accordingly, the ω_z/r ratio increases. Figure 4 summarizes this discussion.

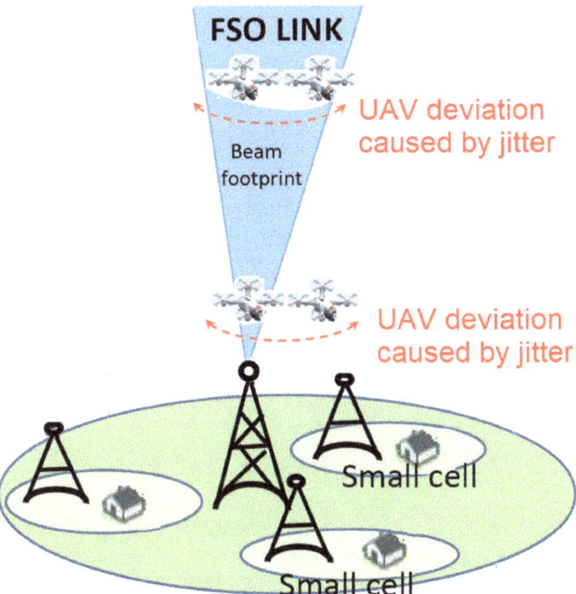

Figure 4. Adverse effect of UAV's jitter caused by its motor vibration versus altitude in the flight state. It is supposed that the same UAV is flying at two different altitudes and affected by a same value of σ_s. For the case of the UAV operating at the lower altitude, the communication link may even be interrupted due to the presence of UAV's jitter. Of course, for a higher altitude, the beam broadening induces more serious power losses at the receiver.

Following the analysis of the channel behavior from Figure 3, now, Figure 5 shows the BER performance of the OOK and OOK-EH schemes under different channel impairments as a function of the received SNR. To illustrate the behavior of the OOK-EH scheme, a $\zeta = 0.8$ has been assumed. In addition, two ratios ω_z/r have been taken as representative values: $(\omega_z/r) = 3$ and $(\omega_z/r) = 10$, corresponding to UAV heights of 300 m and 1000 m, assuming a transmission divergence $\theta_T = 1$ mrad. BER curves plotted in black represent the case of $(\omega_z/r) = 3$; whilst the curves plotted in red depict the performance for $(\omega_z/r) = 10$. In addition, solid lines represent the BER of the ideal channel with $h = A_0$, i.e., assuming only the geometric loss; whereas the dashed lines include the adverse effect of the random medium, i.e., either solely turbulence or both turbulence and pointing error.

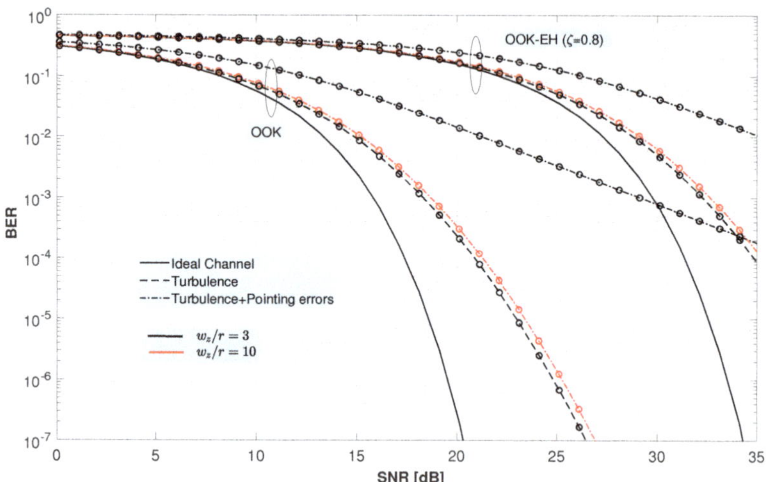

Figure 5. Average BER for the OOK and OOK-EH schemes versus signal-to-noise ratio for a vertical FSO link under different channel impairments and different values of the ratio (w_z/r). A transmitter with a divergence $\theta_T = 1$ mrad and a P_m up to 20 dBm is assumed.

These results, which are calculated from the expression (25) for the case of only considering turbulence-induced fading; and from (26) for the combined scenario with turbulence and pointing error, have been validated using Monte Carlo simulations. Note that BER curves for the ideal channel and the one associated to the turbulent channel with no pointing errors are identical for both (ω_z/r) ratios. In this figure, simulation (Monte Carlo) results are drawn with markers whereas theoretical results are drawn with either solid or dashed lines. In all cases, a perfect match is shown between the simulated results and those obtained from the derived expressions.

From the aforementioned Figure 5, it can be seen that turbulence and pointing error affect the BER with different severity depending on the ω_z/r ratio. As far as turbulence is concerned, it affects BER equally regardless of the ratio ω_z/r. However, as expected, pointing error causes a very severe degradation for lower ω_z/r ratios. The figure shows that, for $(\omega_z/r) = 3$, the SNR degradation caused by pointing errors is huge, on the order of 15 dB for a target BER of 5×10^{-5}, while for $(\omega_z/r) = 10$, the degradation is nearly negligible. In fact, the BER curves considering turbulence and turbulence with pointing error almost overlap.

Optimization for EH

As described in the previous section, the process of optimizing the modified OOK scheme for EH consists of choosing the optimal ζ value that maximizes the average EH while keeping the BER below the target value for each channel condition. Note that, since the increase in EH is achieved at the cost of degrading the link quality, the maximum ζ value will depend on the considered channel impairments, i.e., turbulent fading, pointing error, and atmospheric attenuation. Consequently, a lower channel degradation will lead to higher ζ values and, thus, to higher harvested energies. The value of the optimal ζ has been obtained numerically from expression (26) by setting the link parameters and the target BER.

To form a complete picture of the dependence of optimum ζ on the FSO link parameters, Figure 6 shows the optimal value of ζ as a function of transmitter peak power considering different ω_z/r ratios, jitters and target BERs. In all cases, the turbulence fading and the path loss corresponding to "very clear air" are included. As can be seen in the figure, as the peak power of the transmitter increases, the optimum ζ value also increases. In our analysis, realistic power values up to 35 dBm have been considered. Note that, since the UAV is intended to collect as much energy as possible, the power value chosen in the link design will be high. Therefore, the optimal ζ values will also be high. For the highest of the powers considered ($P_m =35$ dBm), the value of ζ is always higher than 0.8. In addition, as expected, the optimal ζ values obtained for the 10^{-8} target rate (red lines) are lower than for 5×10^{-5} (blue lines) and, similarly, higher jitter values lead to lower ζ values. From that Figure 6, it can be concluded that values of ζ higher than 0.8 one can be selected for realistic propagation scenarios.

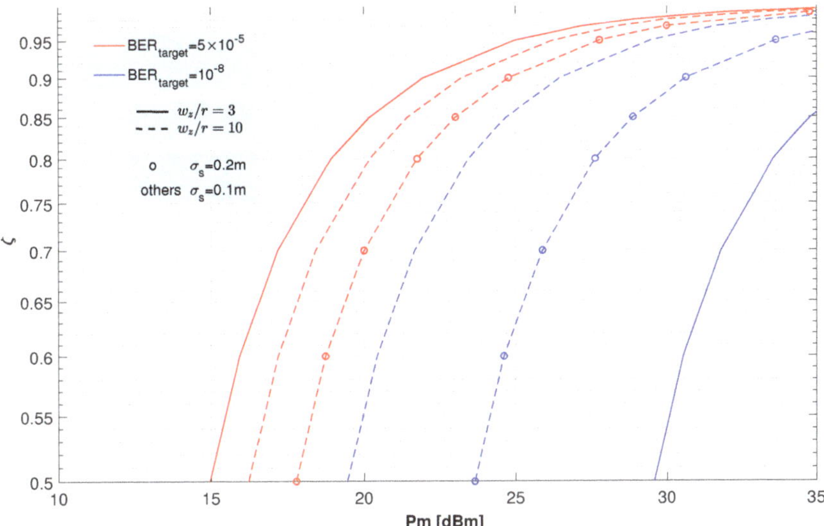

Figure 6. Optimal ζ values for the proposed OOK-EH scheme as a function of the transmitted peak power considering different (ω_z/r) ratios, jitters and target BER. These ζ values maximize the average harvested energy while maintaining the target BER. Turbulence fading ($\alpha = 30$, $\beta = 30$) and "very clear air" conditions are assumed.

On a different matter, Figure 7 depicts the average harvested energy for the OOK and OOK-EH schemes as a function of the peak transmitted power. For the OOK-EH scheme, the optimal ζ values obtained in Figure 6 have been here employed.

Hence, from the AEH results depicted in Figure 7, the following comments can be drawn. First, it is clearly observed that the energy collected with the OOK-EH scheme (black lines) is higher than that of the OOK scheme (red lines). In fact, the energy harvested by the OOK-EH scheme tends to twice that in OOK scheme as the transmitted power increases. Figure 7 also shows that a minimum P_m is required to achieve the previously selected BER target (a power below P_m cannot satisfy the performance required by the BER target and, accordingly, no energy would be harvested). This P_m is higher for the ratio $\omega_z/r =3$ than for $\omega_z/r =10$ due to what was explained when introducing Figures 3–6. Therefore, the choice of the ratio ω_z/r has a huge impact on the value of the collected energy. The figure shows that the energy obtained with $\omega_z/r = 3$ is much higher than that obtained with $\omega_z/r = 10$. In particular, for the highest power considered (Pm=35 dB), the AEH is about 70 mJ/s for $\omega_z/r = 3$, while it hardly reaches 10 mJ/s for $\omega_z/r = 10$. Note that the AEH values shown in that figure are consistent with those ones published by other authors in [31] and [47] for peak transmitted powers of 100 and 200 mW, respectively.

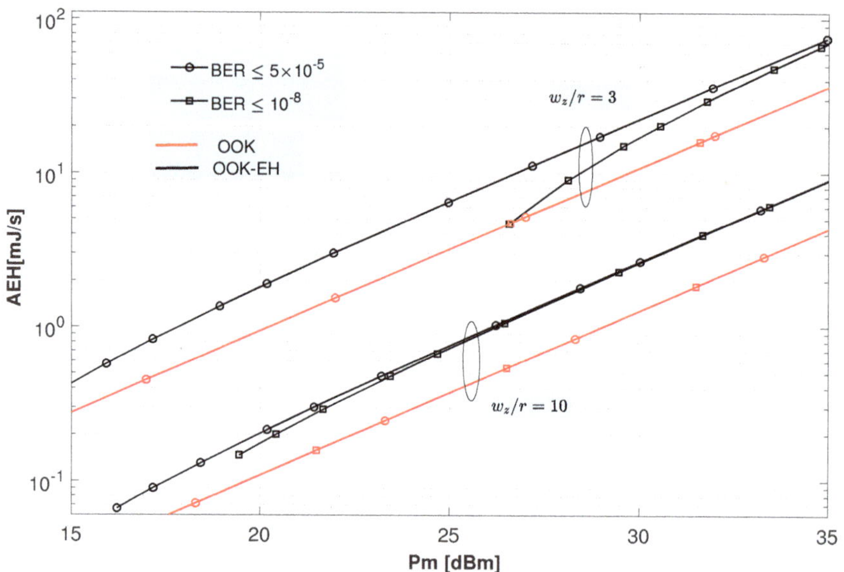

Figure 7. Average harvested energy for the OOK and OOK-EH schemes versus peak transmitted power considering different (ω_z/r) ratios and target BER. Turbulence fading (α =30, β =30), a jitter ($\sigma_s = 0.1$ m) and "very clear air" conditions are assumed. A single transmitter has been considered.

As explained above, one of the main features affecting energy harvesting is the ω_z/r ratio. In this respect, it is possible to design transmitters for EH with an accurate control of their beam width since significant gains can be achieved with the appropriate selection of ω_z/r. Thus, when increasing this ratio, the pointing problem is relaxed and, as well, the negative effect inherent to the jitter is reduced, i.e., when increasing ω_z/r then fading induced by jitter suffered by the receiver is less intense since the received beam waist becomes (much) bigger than the receiver aperture radius. In this respect, although the received spot can wander due to the jitter effect, however, the total amount of caught power in the receiver side is maintained without significant variations since the size of the received beam waist is large compared to the receiver aperture radius. Consequently, the amount of power that can be captured from the section of the beam illuminating the receiver is, more or less, of the same magnitude. On the contrary, for smaller values of ω_z/r, the sway in the received beam footprint caused by misalignment is more critical in the sense that such a sway may make all the received beam spot drop out of the receiver photosensitive area. For that critical situation, the amount of captured power in the receiver would drop to zero. Of course, large ω_z/r values lead to both severe geometrical losses (since most of the received footprint area is spread out of the physical photosensitive area implemented in the receiver side) and, consequently, low values of harvested energy. Therefore, the design of any FSO link should try to minimize this ratio by using adaptive pointing tracking systems [18].

It is worth noting that all the the results of average EH shown so far are for a single vertical FSO link between a given BS and UAV. Nevertheless, each UAV could receive simultaneous transmissions from a number of BSs as long as their locations fall within the region, $\mathfrak{R} \subset \mathbb{R}^2$, covered by the FOV of the UAV's receiver. The area of such a region, $|\mathfrak{R}|$, can be expressed in terms of the FOV angle and the height of the UAV, Z, as follows:

$|\mathfrak{R}| = \pi(Z\tan(\text{FOV}/2))^2$. Therefore, the average number of BSs, N, that are covered by the FOV of the receiver is written as $N = |\mathfrak{R}|\lambda_{\text{BS}}$, where λ_{BS} is the BS density expressed in number of nodes per m^2. Assuming that the macro BSs are spatially distributed according to a hexagonal grid, with a number of RRHs, n_{RRHs}, randomly distributed within each macro cell as described in [59], that BS density leads to $\lambda_{\text{BS}} = \frac{(1+n_{\text{RRHs}})}{2\sqrt{3}\left(\frac{\text{ISD}}{2}\right)^2}$, where the term ISD stands for the inter-site distance (ISD), which is the distance between two neighbouring macro BSs. Finally, according to sections 6.1.2 and 6.1.4 of [59], Thus, for a given FOV, the value of N can be obtained as a function of the UAV height. In the case of dense urban microcellular deployment scenarios, the value of N increases from 6 to 62 when the height of the UAV increases from 300 m to 1000 m. However, despite the significant increase in the number of BS, the total collected energy hardly changes due to the increase of optical link loss with UAV height.

As a conclusion, considering the AEH results shown in Figure 7 and the number of BSs covered by the UAV's receiver described above, the total AEH that can be obtained for considered scenario in very clear air conditions with a target BER of 5×10^{-5} and assuming realistic transmitted powers between 30 and 35 dBm is in the range between 134 and 450 mJ/s for a UAV height of 300 m and between 164 and 558 mJ/s for a UAV height of 1000 m. This free energy collected from the information-carrying FSO signals complements the energy harvested from the terrestrial optical and RF wireless power transfer (WPT) charging stations and contributes to extend the battery life of UAVs.

5. Concluding Remarks

In this paper, we have investigated the application of FSO, UAVs, and EH as an adaptable and efficient solution to provide backhaul/fronthaul connectivity to 5G+ networks. There are many benefits supporting this approach. First, FSO technology provides high bandwidth for the 5G RAN. Second, FSO communication links do not interfere with the RF based 5G RAN. Third, the UAVs, which act as flying nodes of a backhaul/fronthaul network, can adapt to changes in weather and traffic conditions to provide reliable links. Nevertheless, the limited battery of the UAVs causes service interruptions when the UAVs need to recharge them. For this reason, we propose the use of EH to collect energy from the transmission of information signals to combine with the EH from the terrestrial optical and RF WPT charging stations. All these techniques are thought to enhance the on-board battery lifetime of UAVs. To assess the benefits of the proposed approach we have considered a realistic yet tractable channel model that includes the effect of turbulence fading, pointing errors and atmospheric attenuation. Using this model, analytical closed-form expressions of the average harvested energy and the bit error rate of an OOK scheme optimized for information transmission and power transfer are derived. The derived expressions allow to evaluate the performance of vertical FSO links between ground-based BSs and UAVs and to properly select the link parameter to optimize the harvested energy while guaranteeing a reliable connection.

Results show that there exists an interesting trade-off between reliability and harvested energy.

Author Contributions: M.C.-V. and A.J.-N. defined the scope of the review manuscript. All the authors compiled the necessary information to elaborate this review manuscript. All the authors derived the expressions and obtained the results. C.Á.-R., M.Á.-R., F.J.M.-V., M.C.-V., T.R. and A.J.-N. wrote the article. All the authors reviewed and edited the manuscript. All authors have read and agreed to the published version of the manuscript.

Funding: This work has been funded by the University of Málaga.

Institutional Review Board Statement: Not applicable.

Informed Consent Statement: Not applicable.

Data Availability Statement: Not applicable.

Conflicts of Interest: The authors declare no conflict of interest.

Abbreviations

The following abbreviations are used in this manuscript

AWGN	Additive white Gaussian noise
BBU	Base band unit
BER	Bit error rate
BS	Base station
C-RAN	Cloud RAN
EH	Energy harvesting
FEC	Forward error correction
FSO	Free space optical
IM/DD	Intensity-modulation direct-detection
MC	Monte Carlo
NFP	Networked flying platform
OOK	On-off keying
pdf	Probability density function
PAM	Pulse amplitude modulation
PPM	Pulse position modulation
RAN	Radio access network
RRH	Radio remote head
RV	Random variable
SLIPT	Simultaneous light-wave information and power transfer
UAV	Unmanned aerial vehicle
V2X	Vehicular-to-everything

References

1. Khalighi, M.A.; Uysal, M. Survey on Free Space Optical Communication: A Communication Theory Perspective. *IEEE Commun. Surv. Tutor.* **2014**, *16*, 2231–2258. [CrossRef]
2. Hamza, A.S.; Deogun, J.S.; Alexander, D.R. Classification Framework for Free Space Optical Communication Links and Systems. *IEEE Commun. Surv. Tutor.* **2019**, *21*, 1346–1382. [CrossRef]
3. Kahn, J.; Barry, J. Wireless infrared communications. *Proc. IEEE* **1997**, *85*, 265–298. [CrossRef]
4. Ghassemlooy, Z.; Arnon, S.; Uysal, M.; Xu, Z.; Cheng, J. Emerging Optical Wireless Communications-Advances and Challenges. *IEEE J. Select. Areas Commun.* **2015**, *33*, 1738–1749. [CrossRef]
5. Wu, S.; Wang, H.; Youn, C.H. Visible light communications for 5G wireless networking systems: from fixed to mobile communications. *IEEE Netw.* **2014**, *28*, 41–45. [CrossRef]
6. Alzenad, M.; Shakir, M.Z.; Yanikomeroglu, H.; Alouini, M.S. FSO-Based Vertical Backhaul/Fronthaul Framework for 5G+ Wireless Networks. *IEEE Commun. Mag.* **2018**, *56*, 218–224. [CrossRef]
7. Technology Demonstration Missions: Laser Communications Relay Demonstration (LCRD). Available online: http://www.nasa.gov/mission_pages/tdm/lcrd/index.html (accessed on 27 April 2022).
8. Odeyemi, K.O.; Owolawi, P.A. A Mixed FSO/RF Integrated Satellite-High Altitude Platform Relaying Networks for Multiple Terrestrial Users with Presence of Eavesdropper: A Secrecy Performance. *Photonics* **2022**, *9*, 32. [CrossRef]
9. Fidler, F. ; Knapek, M.; Horwath, J.; Leeb, W.R. Optical Communications for High-Altitude Platforms. *IEEE J. Select. Top. Quant. Electron.* **2010**, *16*, 1058–1070. [CrossRef]
10. Tang, S.; Dong, Y.; Zhang, X. Impulse Response Modeling for Underwater Wireless Optical Communication Links. *IEEE Trans. Commun.* **2014**, *62*, 226–234. [CrossRef]
11. Jurado-Navas, A.; González Serrato, N.; Garrido-Balsells, J.M.; Castillo-Vázquez, M. Error probability analysis of OOK and variable weight MPPM coding schemes for underwater optical communication systems affected by salinity turbulence. *OSA Cont.* **2018**, *1*, 1131–1143. [CrossRef]
12. Jurado-Navas, A.; Álvarez Roa, C.; Álvarez Roa, M.; Castillo-Vázquez, M. Cooperative Terrestrial-Underwater Wireless Optical Links by Using an Amplify-and-Forward Strategy. *Sensors* **2022**, *22*, 2464. [CrossRef] [PubMed]
13. Jurado-Navas, A.; Garrido-Balsells, J.M.; Castillo-Vázquez, M.; García-Zambrana, A.; Puerta-Notario, A. Converging Underwater and FSO Ground Communication Links. In Proceedings of the 2019 Optical Fiber Communications Conference and Exhibition (OFC), San Diego, CA, USA, 3–7 March 2019; pp. 1–3. (Invited paper).
14. Chowdhury, M.I.S.; Kavehrad, M.; Zhang, W.; Deng, P. Combined CATV and Very-High-Speed Data Transmission over a 1550-nm Wavelength Indoor Optical Wireless Link. In *Next-Generation Optical Networks for Data Centers and Short-Reach Links*; Srivastava, A.K., Ed.; International Society for Optics and Photonics, SPIE: Bellingham, WA, USA, 2014; Volume 9010, pp. 38–45. [CrossRef]

15. Hamza, A.S.; Deogun, J.S.; Alexander, D.R. Wireless Communication in Data Centers: A Survey. *IEEE Commun. Surv. Tutor.* **2016**, *18*, 1572–1595. [CrossRef]

16. Raddo, T.R.; Perez-Santacruz, J.; Johannsen, U.; Dayoub, I.; Haxha, S.; Monroy, I.T.; Jurado-Navas, A. FSO-CDMA Systems Supporting end-to-end Network Slicing. In Proceedings of the Imaging and Applied Optics Congress, Washington, DC, USA, 22–26 June 2020; Optical Society of America: Washington, DC, USA, 2020; p. JW2A.38.

17. Qin, Y.; Kishk, M.A.; Alouini, M.S. Drone Charging Stations Deployment in Rural Areas for Better Wireless Coverage: Challenges and Solutions. *IEEE Internet Things Mag.* **2022**, *5*, 148–153. [CrossRef]

18. Kaymak, Y.; Rojas-Cessa, R.; Feng, J.; Ansari, N.; Zhou, M.; Zhang, T. A Survey on Acquisition, Tracking, and Pointing Mechanisms for Mobile Free-Space Optical Communications. *IEEE Commun. Surv. Tutor.* **2018**, *20*, 1104–1123. [CrossRef]

19. Andrews, L.; Phillips, R.; Hopen, C.; Al-Habash, M. Theory of optical scintillation. *J. Opt. Soc. Am. A* **1999**, *16*, 1417–1429. [CrossRef]

20. Strohbehn, J. Modern theories in the propagation of optical waves in a turbulent medium. In *Laser Beam Propagation in the Atmosphere*; Springer: Berlin/Heidelberg, Germany, 1978; pp. 45–106.

21. Al-Habash, M.; Adrews, L.; Phillips, R. Mathematical model for the irradiance probability density function of a laser beam propagating through turbulent media. *Opt. Eng.* **2001**, *40*, 1554–1562. [CrossRef]

22. Barrett, J.L.; Budni, P.A. Laser beam propagation through strong turbulence. *J. Appl. Phys.* **1992**, *71*, 1124–1127. [CrossRef]

23. Khalighi, M.; Xu, F.; Jaafar, Y.; Bourennane, S. Double-Laser Differential Signaling for Reducing the Effect of Background Radiation in Free-Space Optical Systems. *J. Opt. Commun. Netw.* **2011**, *3*, 145–154. [CrossRef]

24. Saleh, B.E.A.; Teich, M.C. *Fundamentals of Photonics*; Wiley: Hoboken, NJ, USA, 1991.

25. Santacruz, J.P.; Rommel, S.; Johannsen, U.; Jurado-Navas, A.; Monroy, I.T. Analysis and Compensation of Phase Noise in Mm-Wave OFDM ARoF Systems for Beyond 5G. *J. Lightwave Technol.* **2021**, *39*, 1602–1610. [CrossRef]

26. Song, S.; Choi, M.; Ko, D.E.; Chung, J.M. Multi-UAV Trajectory Optimization Considering Collisions in FSO Communication Networks. *IEEE J. Select. Areas Commun.* **2021**, *39*, 3378–3394. [CrossRef]

27. Lee, J.H.; Park, K.H.; Alouini, M.S.; Ko, Y.C. Free Space Optical Communication on UAV-Assisted Backhaul Networks: Optimization for Service Time. In Proceedings of the 2019 IEEE Globecom Workshops (GC Wkshps), Waikoloa, HI, USA, 9–13 December 2019; pp. 1–6. [CrossRef]

28. Ulukus, S.; Yener, A.; Erkip, E.; Simeone, O.; Zorzi, M.; Grover, P.; Huang, K. Energy Harvesting Wireless Communications: A Review of Recent Advances. *IEEE J. Select. Areas Commun.* **2015**, *33*, 360–381. [CrossRef]

29. Diamantoulakis, P.D.; Karagiannidis, G.K.; Ding, Z. Simultaneous Lightwave Information and Power Transfer (SLIPT). *IEEE Trans. Green Commun. Netw.* **2018**, *2*, 764–773. [CrossRef]

30. Chen, J.; Yang, L.; Wang, W.; Yang, H.C.; Liu, Y.; Hasna, M.O.; Alouini, M.S. A Novel Energy Harvesting Scheme for Mixed FSO-RF Relaying Systems. *IEEE Trans. Vehic. Technol.* **2019**, *68*, 8259–8263. [CrossRef]

31. Abou-Rjeily, C.; Kaddoum, G.; Karagiannidis, G.K. Ground-to-air FSO communications: when high data rate communication meets efficient energy harvesting with simple designs. *Opt. Express* **2019**, *27*, 34079–34092. [CrossRef] [PubMed]

32. Lahmeri, M.A.; Kishk, M.A.; Alouini, M.S. Stochastic Geometry-Based Analysis of Airborne Base Stations With Laser-Powered UAVs. *IEEE Commun. Lett.* **2020**, *24*, 173–177. [CrossRef]

33. Che, Y.L.; Long, W.; Luo, S.; Wu, K.; Zhang, R. Energy-Efficient UAV Multicasting With Simultaneous FSO Backhaul and Power Transfer. *IEEE Wireless Commun. Lett.* **2021**, *10*, 1537–1541. [CrossRef]

34. Shehzad, M.K.; Ahmad, A.; Hassan, S.A.; Jung, H. Backhaul-Aware Intelligent Positioning of UAVs and Association of Terrestrial Base Stations for Fronthaul Connectivity. *IEEE Trans. Netw. Sci. Eng.* **2021**, *8*, 2742–2755. [CrossRef]

35. Gonzalez-Diaz, S.; Garcia-Saavedra, A.; de la Oliva, A.; Costa-Perez, X.; Gazda, R.; Mourad, A.; Deiss, T.; Mangues-Bafalluy, J.; Iovanna, P.; Stracca, S.; et al. Integrating Fronthaul and Backhaul Networks: Transport Challenges and Feasibility Results. *IEEE Trans. Mob. Comput.* **2021**, *20*, 533–549. [CrossRef]

36. Jaber, M.; Imran, M.A.; Tafazolli, R.; Tukmanov, A. 5G Backhaul Challenges and Emerging Research Directions: A Survey. *IEEE Access* **2016**, *4*, 1743–1766. [CrossRef]

37. Zhang, Z.; Xiao, Y.; Ma, Z.; Xiao, M.; Ding, Z.; Lei, X.; Karagiannidis, G.K.; Fan, P. 6G Wireless Networks: Vision, Requirements, Architecture, and Key Technologies. *IEEE Veh. Technol. Mag.* **2019**, *14*, 28–41. [CrossRef]

38. Álvarez Roa, M.; Álvarez Roa, C.; Fernández-Aragón, F.; Raddo, T.; Garrido-Balsells, J.M.; Monroy, I.T.; Jurado-Navas, A. Performance analysis of atmospheric optical communication systems with spatial diversity affected by correlated turbulence. *J. Opt. Commun. Netw.* **2022**, *14*, 524–539. [CrossRef]

39. Andrews, L.C.; Phillips, R.L. *Laser Beam Propagation Through Random Media*; SPIE: Bellingham, WA, USA, 2005.

40. Jurado-Navas, A.; Garrido-Balsells, J.M.; Paris, J.F.; Castillo-Vázquez, M.; Puerta-Notario, A. Impact of pointing errors on the performance of generalized atmospheric optical channels. *Opt. Express* **2012**, *20*, 12550–12562. [CrossRef] [PubMed]

41. Garrido-Balsells, J.M.; Lopez-Martinez, F.J.; Castillo-Vázquez, M.; Jurado-Navas, A.; Puerta-Notario, A. Performance analysis of FSO communications under LOS blockage. *Opt. Express* **2017**, *25*, 25278–25294. [CrossRef]

42. Weichel, H. *Laser Beam Propagation in the Atmosphere*; SPIE Optical Engineering Press: Bellingham, WA, USA, 1990.

43. Naboulsi, A.; Sizun.; Fornel, D. Propagation of optical and infrared waves in the atmosphere. In Proceedings of the XXVIIIth URSI General Assembly, New Delhi, India, 23–29 October 2005.

44. Farid, A.A.; Hranilovic, S. Outage Capacity Optimization for Free-Space Optical Links With Pointing Errors. *J. Lightwave Technol.* **2007**, *25*, 1702–1710. [CrossRef]
45. Farid, A.A.; Hranilovic, S. Outage Probability for Free-Space Optical Systems Over Slow Fading Channels With Pointing Errors. In Proceedings of the LEOS 2006—19th Annual Meeting of the IEEE Lasers and Electro-Optics Society, Montreal, QC, Canada, 29 October–2 November 2006; pp. 82–83. [CrossRef]
46. Zhou, X.; Zhang, R.; Ho, C.K. Wireless Information and Power Transfer: Architecture Design and Rate-Energy Tradeoff. *IEEE Trans. Commun.* **2013**, *61*, 4754–4767. [CrossRef]
47. Abou-Rjeily, C.; Kaddoum, G. Free Space Optical Cooperative Communications via an Energy Harvesting Harvest-Store-Use Relay. *IEEE Trans. Wireless Commun.* **2020**, *19*, 6564–6577. [CrossRef]
48. Wolfram. Available online: http://functions.wolfram.com/ (accessed on 24 July 2022).
49. IEEE Standard for Ethernet–Amendment 2: Media Access Control Parameters, Physical Layers, and Management Parameters for 25 Gb/s Operation Amendment 2: Media Access Control Parameters, Physical Layers, and Management Parameters for 25 Gb/s Operation. In *IEEE Std 802.3by-2016 (Amendment to IEEE Std 802.3-2015 as amended by IEEE Std 802.3bw-2015)*; IEEE: Piscataway, NJ, USA, 2016; pp. 1–244. [CrossRef]
50. Jurado-Navas, A.; Garrido-Balsells, J.M.; Castillo-Vázquez, M.; Puerta-Notario, A.; Tafur-Monroy, I.; Vegas-Olmos, J.J. Optimal threshold detection for Málaga turbulent optical links. *Opt. Appl.* **2016**, *46*, 577–595.
51. Song, T.; Wang, Q.; Wu, M.W.; Ohtsuki, T.; Gurusamy, M.; Kam, P.Y. Impact of Pointing Errors on the Error Performance of Intersatellite Laser Communications. *J. Lightwave Technol.* **2017**, *35*, 3082–3091. [CrossRef]
52. Jurado-Navas, A.; Garrido-Balsells, J.M.; Paris, J.F.; Puerta-Notario, A. A unifying statistical model for atmospheric optical scintillation. In *Numerical Simulations of Physical and Engineering Processes*; Awrejcewicz, J., Ed.; In-Tech: Rijeka, Croatia, 2011; pp. 181–206.
53. Garrido-Balsells, J.M.; Jurado-Navas, A.; Paris, J.F.; Castillo-Vazquez, M.; Puerta-Notario, A. Novel formulation of the $\mathcal{M}$ model through the Generalized-K distribution for atmospheric optical channels. *Opt. Express* **2015**, *23*, 6345–6358. [CrossRef]
54. Wang, Q.; Lin, H.; Kam, P.Y. Tight bounds and invertible average error probability expressions over composite fading channels. *J. Commun. Netw.* **2016**, *18*, 182–189. [CrossRef]
55. EKSMA Optics. Available online: https://eksmaoptics.com/ (accessed on 24 July 2022).
56. Edmun Optics. 1550nm Laser Line Coated Fused Silica PCX Lenses. Available online: https://www.edmundoptics.es/f/1550nm-laser-line-coated-fused-silica-pcx-lenses/15064/ (accessed on 24 July 2022).
57. 3GPP. *Technical Specification Group Radio Access Network: Study on Enhanced LTE Support for Aerial Vehicles*; Technical Report (TR) 36.777, 3rd Generation Partnership Project (3GPP); (3GPP) Mobile Competence Centre: Sophia Antipolis, France, 2017; Release 15.0.0.
58. Wang, J.Y.; Ma, Y.; Lu, R.R.; Wang, J.B.; Lin, M.; Cheng, J. Hovering UAV-Based FSO Communications: Channel Modelling, Performance Analysis, and Parameter Optimization. *IEEE J. Select. Areas Commun.* **2021**, *39*, 2946–2959. [CrossRef]
59. 3GPP. *5G; Study on Scenarios and Requirements for Next Generation Access Technologies*; Technical Report (TR) 39.913, 3rd Generation Partnership Project (3GPP); (3GPP) Mobile Competence Centre: Sophia Antipolis, France, 2022; Version 17.0.0.

Article

Matched Filtering for MIMO Coherent Optical Communications with Mode-Dependent Loss Channels

Luis M. Torres [1,*,†], Francisco J. Cañete [2] and Luis Díez [2]

1   Research and Development Department, KDPOF SL, 28760 Madrid, Spain
2   Communications and Signal Processing Lab, Instituto Universitario de Investigación en Telecomunicación (TELMA), ETSI Telecomunicación Universidad de Málaga, 29071 Málaga, Spain; francis@ic.uma.es (F.J.C.); diez@ic.uma.es (L.D.)
*   Correspondence: luismanuel.torres@gmail.com
†   Current address: Ronda de Poniente 14, 2CD, 28760 Madrid, Spain.

Abstract: The use of digital signal processors (DSP) to equalize coherent optical communication systems based on spatial division multiplexing (SDM) techniques is widespread in current optical receivers. However, most of DSP implementation approaches found in the literature assume a negligible mode-dependent loss (MDL). This paper is focused on the linear multiple-input multiple-output (MIMO) receiver designed to optimize the minimum mean square error (MMSE) for a coherent SDM optical communication system, without previous assumptions on receiver oversampling or analog front-end realizations. The influence of the roll-off factor of a generic pulse-amplitude modulation (PAM) transmitter on system performance is studied as well. As a main result of the proposed approach, the ability of a simple match filter (MF) based MIMO receiver to completely eliminate inter-symbol interference (ISI) and crosstalk for SDM systems under the assumption of negligible MDL is demonstrated. The performance of the linear MIMO fractionally-spaced equalizer (FSE) receiver for an SDM system with a MDL-impaired channel is then evaluated by numerical simulations using novel system performance indicators, in the form of signal to noise and distortion ratio (SNDR) loss, with respect to the case without MDL. System performance improvements by increasing the transmitter roll-off factor are also quantified.

Keywords: coherent optical communication; optical fiber communication; MIMO adaptive equalizer; matched filter; MMSE; spatial division multiplexing (SDM); polarization division multiplexing; fractional-spaced equalizer (FSE)

Citation: Torres, L.M.; Cañete, F.J.; Díez, L. Matched Filtering for MIMO Coherent Optical Communications with Mode-Dependent Loss Channels. *Sensors* **2022**, *22*, 798. https://doi.org/10.3390/s22030798

Academic Editors: Jiangbing Du, Yang Yue, Jian Zhao and Yan-ge Liu

Received: 10 December 2021
Accepted: 18 January 2022
Published: 21 January 2022

Publisher's Note: MDPI stays neutral with regard to jurisdictional claims in published maps and institutional affiliations.

1. Introduction

The increasing demand of higher bit rates, combined with the environmental requirement of energy-efficient communication systems, is driving the development of ultra-high-capacity fiber optic communications. In this context, recent advances in spatial division multiplexing (SDM) using multimode or multicore fibers in long- and short-distance links [1,2] cannot be possible without the extensive use of multiple-input multiple-output (MIMO) signal processing.

Since the initial proposal to use polarization-division multiplexing (PDM) in a single mode fiber (SMF) [3] to double the capacity of a coherent optical communication system, MIMO signal processing [4,5] has become necessary to process and recover the parallel transmitted data streams even before the signal processing used was identified as a MIMO equalizer [6]. The channel model for PDM in SMF and its relation with the non-linear Schrödinger equation [7], its representation by means of the 2 × 2 Jones matrix [8] and as a multi-section system [9] has been extensively discussed in previous works. Multiple contributions to adaptive MIMO equalizers using the flexibility of digital signal processors (DSP) have been developed [7,10,11], where normally the equalization is divided into two parts: The first one, with an invariant chromatic dispersion (CD) compensation for each of the polarizations; and a second one, with an adaptive 2×2 MIMO linear equalizer to resolve the crosstalk between the modes [12].

SDM [13] using multimode fibers (MMF) [14] or few mode fibers (FMF) [15,16], appeared as a solution for communication systems reaching speeds well above 100 Tb/s when combined with wavelength multiplexing techniques [17]. Therefore, the optical channel model, based on the Jones matrix, was extended to represent the multiple fiber modes [18–20], and adaptive linear MIMO equalization [21] was studied and updated as an extension of the PDM case [2]. There are works that study the complexity of direct time and frequency domain implementations of the adaptive MIMO equalization, both for a linear design [12,22–24] and a nonlinear one [25], and also in the optical domain [26]. An important difference of SDM systems w.r.t. PDM systems is that the modal dispersion (MD) in SDM systems is higher than the equivalent polarization dispersion in PDM systems, reaching the same order of magnitude of the CD [20]. This boosts looking for simpler DSP schemes that avoid the enormous complexity required from the classical equalizers proposed for a PDM system and initially adapted to SDM systems [24]. In particular, linear MIMO receiver designs have been proposed for SDM systems [27,28] by expanding PDM systems [12,21,29], where a fractional-spaced equalizer (FSE) with an oversampling rate r_{ov} of two is used. A review of different combinations of fiber types and DSP schemes reported in the literature with their associated complexity is summarized in [1].

The impact of mode-dependent loss (MDL) in long-haul optical links has more recently been studied, especially in the associated loss in the channel capacity when using minimum mean square error (MMSE) MIMO receivers [27,30]. This fact has initiated a race towards nonlinear receivers that can improve performance in the presence of MDL, increasing the receiver complexity notably [31,32]. However, performance evaluation of SDM systems that incorporate MIMO FSE receivers in the presence of MDL and the impact that pulse-amplitude modulation (PAM) pulses roll-off factor have on this performance, still deserve attention.

This paper provides a framework for the analysis of linear MIMO receivers for SDM that includes a continuous-time MIMO matched filter followed by a MIMO linear filter, without making prior assumptions about oversampling or the continuous-time optical front-end. This approach provides, for example, a theoretical basis for possible silicon photonics optical front-ends capable of SDM equalization. We show that the generalized linear MIMO MMSE receiver, for channels with negligible MDL, can be simplified to a matched filter MIMO receiver, which completely eliminates the ISI and crosstalk introduced by the channel.

When the optical channel exhibits a significant MDL, we include linear equalization and carry out numerical simulations to get the performance of a system that consists of: A PAM transmitter with square-root raised cosine pulses; a complete long-haul optical channel with SDM; and a MIMO receiver based on the FSE approach with oversampling of two. To this end, an ensemble of thousands of random optical channels has been generated and the system performance is evaluated by means of the signal to noise and distortion ratio (SNDR) loss at the receiver output w.r.t., the one of an optimal equivalent system without ISI and crosstalk. These results are presented for a configuration with a set of parameters for a fiber, transmitter, and receiver, which is representative of current technology.

The paper is structured as follows. After a short section of defining the notation used (Section 2), we begin by describing the optical channel model for a long-haul communication system using SDM, including CD, MD, and MDL impairments (Section 3). Next, a communication system based on a generalized PAM transmitter with square-root raised cosine pulses, and a linear MIMO receiver designed under the MMSE optimization criterion are discussed in Section 4. In Section 5, the numerical simulations are presented and SNDR loss metrics are given for the optical communication system with a FSE MIMO receiver for different values of channel dispersion (including MDL) and roll-off factor of the square-root raised cosine pulses. Finally, conclusions are summarized in Section 6.

2. Notation

Matrices are represented as $\underline{M}$, and vectors as $\underline{v}$. Vectors are column vectors unless otherwise noted. $\lfloor x \rfloor$ denotes the largest integer less than or equal to x. x^* represents the conjugate of x, $\underline{M}^H$ denotes the Hermitian of $\underline{M}$, and $\underline{v}^T$ represents the transpose of the vector $\underline{v}$. $E[x]$ is the expectation operator applied to the random variable x. $i \in \{1, \ldots, D\}$ is used to index a mode among the D modes used in the fiber and $*$ represents the convolution operator. The result of the convolution operator applied to a $D_1 \times D_2$ matrix $\underline{a}(t)$ and a $D_2 \times D_3$ matrix $\underline{b}(t)$ is a $D_1 \times D_3$ matrix denoted as $\underline{c}(t)$ and given by:

$$\underline{c}(t) = \underline{a}(t) * \underline{b}(t) \tag{1}$$

where each of the elements of $\underline{c}(t)$, denoted as $c_{ij}(t)$, are obtained as in a simple matrix multiplication, but substituting the product by the convolution operator:

$$c_{ij}(t) = \sum_{k=1}^{D_2} a_{ik}(t) * b_{kj}(t). \tag{2}$$

Similarly, the result of the convolution operator applied to a $D_1 \times D_2$ matrix $\underline{a}(t)$ and a time dependent signal $y(t)$ is a $D_1 \times D_2$ matrix denoted as $\underline{d}(t)$ where $\underline{d}(t) = \underline{a}(t) * y(t)$. Each of the elements of the matrix $\underline{d}(t)$ are obtained as in a multiplication of a matrix with a scalar, however substituting the product by the convolution operator:

$$d_{ij}(t) = a_{ij}(t) * y(t). \tag{3}$$

$\mathcal{F}\{y(t)\}$ denotes the Fourier transform of the continuous-time signal $y(t)$ and $\mathcal{F}^{-1}\{Y(\omega))\}$ the denotes inverse Fourier transform of $Y(\omega)$. Similarly, for the discrete-time signal $y[n]$ we denote its corresponding discrete Fourier transform as $Y(\Omega)$.

3. Long-Haul Optical Link MIMO Channel Model

In this section we describe the multi-section optical channel model used in this work. The effect of the channel noise is discussed separately in Section 4. The relationship between the input vector $\underline{x}(\omega) = [x_1(\omega), x_2(\omega), \ldots, x_D(\omega)]^T$ of complex electric field amplitudes of each of the D modes propagating along the fiber, and the corresponding output vector $\underline{y}(\omega) = [y_1(\omega), y_2(\omega), \ldots, y_D(\omega)]^T$ can be modeled, after neglecting non-linear effects, as a multiple-input multiple-output linear system $\underline{\underline{H}}_{tot}(\omega)$ [2]:

$$\underline{y}(\omega) = \underline{\underline{H}}_{tot}(\omega)\underline{x}(\omega), \tag{4}$$

where $\underline{\underline{H}}_{tot}(\omega)$ is a $D \times D$ matrix that models the signal propagation along the channel. For $D = 2$, the system is equivalent to a classical PDM over a SMF, and $\underline{\underline{H}}_{tot}(\omega)$ takes the form of the Jones matrix [8]. For $D > 2$, extensions to the Jones matrix have been proposed to be adequate for the SDM model [19,20].

In the case of long-haul systems, $\underline{\underline{H}}_{tot}(\omega)$ can be further modeled as a concatenation of K_{amp} spans composed of the optical fiber and an optical amplifier [2,18,33,34]. Hence, the whole channel transfer function can be written as:

$$\underline{\underline{H}}_{tot}(\omega) = H_{CD}(\omega) \cdot \underline{\underline{H}}(\omega), \tag{5}$$

where $H_{CD}(\omega) = e^{\left(-\frac{j}{2}\omega^2 \bar{\beta}_2 \ell_{tot}\right)}$ is a single-input single-output (SISO) term that models the mode-averaged distortion due to CD, $\bar{\beta}_2$ represents the mode-averaged CD per unit length, and ℓ_{tot} denotes the total link length. The matrix $\underline{\underline{H}}(\omega)$ includes inter-mode cross-talk, MDL and MD effects of the complete MIMO system. Equation (5) can be written as a product over the K_{amp} spans:

$$\underline{\underline{H}}(\omega) = \prod_{k=1}^{K_{amp}} \underline{\underline{H}}^{(k)}(\omega), \tag{6}$$

where $\underline{\underline{H}}^{(k)}(\omega)$ is the channel response of the kth span. We use $k \in \{1, \ldots, K_{amp}\}$ to index the spans in the optical channel. We can write out $\underline{\underline{H}}^{(k)}(\omega)$ as [20,34]:

$$\underline{\underline{H}}^{(k)}(\omega) = \underline{\underline{V}}^{(k)}\underline{\underline{\Lambda}}^{(k)}(\omega)(\underline{\underline{U}}^{(k)})^{H}, \tag{7}$$

where the diagonal matrix $\underline{\underline{\Lambda}}^{(k)}(\omega)$ for a given span k includes the MDL effects and the MD of each mode w.r.t. the mode-averaged value [33], and can be expressed as:

$$\underline{\underline{\Lambda}}^{(k)}(\omega) = \text{diag}\left(\left[e^{\left(\frac{1}{2}g_1^{(k)} - j\omega\tau_1^{(k)}\right)}, \ldots, e^{\left(\frac{1}{2}g_D^{(k)} - j\omega\tau_D^{(k)}\right)}\right]\right), \tag{8}$$

being $\underline{g}^{(k)} = [g_1^{(k)}, g_2^{(k)}, \ldots, g_D^{(k)}]$ the uncoupled modal gains and $\underline{\tau}^{(k)} = [\tau_1^{(k)}, \tau_2^{(k)}, \ldots, \tau_D^{(k)}]$ uncoupled modal group delays. We assume that the uncoupled modal group-velocity dispersion is equal to zero for all the k spans [33].

The k-th span mode coupling is modeled by the frequency-independent $\underline{\underline{V}}^{(k)}$ and $\underline{\underline{U}}^{(k)}$ matrices. It is important to note that, by considering that all the modes propagating through the fiber experience the same attenuation, both matrices are unitary, i.e.,

$$\underline{\underline{V}}^{(k)} \cdot (\underline{\underline{V}}^{(k)})^{H} = \underline{\underline{I}} = \underline{\underline{U}}^{(k)} \cdot (\underline{\underline{U}}^{(k)})^{H}. \tag{9}$$

Alternatively, $\underline{\underline{H}}(\omega)$ can also be written by applying a singular value decomposition (SVD), as the product of two unitary matrices $\underline{\underline{U}}^{(tot)}(\omega)$ and $\underline{\underline{V}}^{(tot)}(\omega)$, and a diagonal matrix $\underline{\underline{\Lambda}}^{(tot)}(\omega)$ as [35]:

$$\underline{\underline{H}}(\omega) = \underline{\underline{V}}^{(tot)}(\omega)\underline{\underline{\Lambda}}^{(tot)}(\omega)\underline{\underline{U}}^{(tot)^{H}}(\omega) \tag{10}$$

where now, the diagonal matrix $\underline{\underline{\Lambda}}^{(tot)}(\omega)$ is given by:

$$\underline{\underline{\Lambda}}^{(tot)}(\omega) = \text{diag}\left(\left[e^{\left(\frac{1}{2}g_1^{(tot)} - j\omega\tau_1^{(tot)}\right)}, \ldots, e^{\left(\frac{1}{2}g_D^{(tot)} - j\omega\tau_D^{(tot)}\right)}\right]\right) \tag{11}$$

where $\underline{g}^{(tot)} = [g_1^{(tot)}, g_2^{(tot)}, \ldots, g_D^{(tot)}]$ are the coupled modal gains of the overall channel and $\underline{\tau}^{(tot)} = [\tau_1^{(tot)}, \tau_2^{(tot)}, \ldots, \tau_D^{(tot)}]$ denote the coupled modal group delays.

Note that in (10), both $\underline{\underline{U}}^{(tot)}(\omega)$ and $\underline{\underline{V}}^{(tot)}(\omega)$ unitary matrices have in general frequency dependence, in contrast to $\underline{\underline{U}}^{(k)}$ and $\underline{\underline{V}}^{(k)}$ in (7) that have not [27,33].

4. SDM Communication System Model

This section describes the model employed to represent the communication system established over the optical channel with multiple spans. The SVD of the channel in (10) can be useful for designing a transmitter based on a precoding matrix combined with a linear receiver, as used in wireless systems [36]. However, this approach becomes unfeasible for long-haul optical communication systems, since the end-to-end channel side information needed to build the transmitter precoding matrix changes faster than the time needed for the system to collect, send, and process that information [33]. Therefore we focus on a SDM system with no channel side information that uses a linear receiver to cope with the channel impairments as shown in Figure 1 [37].

The binary data symbols, $\underline{s}[n] = [s_1[n], s_2[n], \ldots, s_D[n]]^{T}$, are PAM modulated in parallel for each of the $i \in \{1, \ldots, D\}$ optical modes using the same transmitter pulse $P(\omega)$ to get the PAM signals, denoted by the column vector $\underline{x}(t) = [x_1(t), x_2(t), \ldots, x_D(t)]^{T}$. In Figure 1, T is the transmitted symbol period and the first block represent D parallel

PAM modulators working at a symbol rate (and, hence, it includes the discrete-time to continuous-time conversion). The transmitted signal is distorted by ISI and crosstalk introduced by the MIMO channel, modeled with the $\underline{\underline{H}}_{tot}(\omega)$ matrix described in Section 3. It has been shown that the noise in a MDL-impaired system is additive and spatially white [21]. Therefore, in this work we add before the receiver, as part of the channel, an additive white Gaussian noise (AWGN) vector $\underline{n}(t) = [n_1(t), n_2(t), \ldots, n_D(t)]^T$, which, for a certain mode i, has a variance equal to $\frac{N_0}{2}$.

The resulting continuous-time signal vector $\underline{y}(t) = [y_1(t), y_2(t), \ldots, y_D(t)]^T$ is processed by a MIMO receiver to obtain the estimation of the transmitted symbols $\underline{s}[n]$, denoted as $\underline{\hat{s}}[n] = [\hat{s}_1[n], \hat{s}_2[n], \ldots, \hat{s}_D[n]]^T$. In this work we focus on linear MIMO receivers and so, the estimation part in the receiver is depicted in Figure 1 with a generic linear filter of response $\underline{O}(\omega)$, which is followed by a sampler working at the symbol rate. In the following, we propose linear MIMO receiver structures based on the MMSE criterion of an estimated symbol vector.

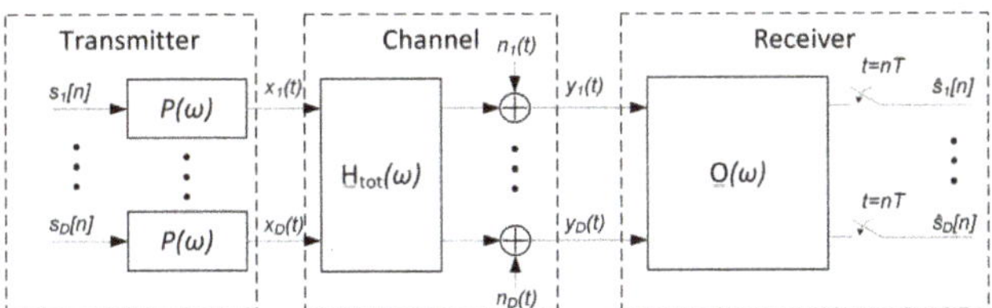

Figure 1. Spatial division multilpexing (SDM) communication system model with linear multiple-input multiple-output (MIMO) receiver.

4.1. Transmitter

The transmitted data in each of the D modes are modulated using a PAM with a square-root raised cosine pulse $p(t)$ with roll-off factor equal to α and normalized power, which can be expressed as [38]:

$$p(t) = \frac{4\alpha}{\pi\sqrt{T}} \cdot \frac{\cos\left([1+\alpha]\frac{\pi t}{T}\right) + \frac{T \cdot \sin\left([1-\alpha]\frac{\pi t}{T}\right)}{4\alpha t}}{1 - \left(\frac{4\alpha t}{T}\right)^2}. \tag{12}$$

Hence, we can write the sequence of PAM pulses for a given mode i as:

$$x_i(t) = \sum_{n=-\infty}^{\infty} s_i[n]p(t - nT), \tag{13}$$

where $s_i[n]$ is a random variable with values taken from the set defined by the PAM modulation scheme. Let us define the global impulse response $\underline{q}(t)$ as the convolution of the transmitting pulse $p(t)$ and the optical channel impulse response matrix $\underline{\underline{h}}_{tot}(t)$ as:

$$\underline{\underline{q}}(t) = \underline{\underline{h}}_{tot}(t) * p(t), \tag{14}$$

so that $\underline{\underline{q}}(t)$ is a $D \times D$ matrix of impulse responses. This way, $q_{ij}(t)$ describes the impulse response between the transmitter mode i and the receiver mode j. Therefore, we can write the relationship between the transmitted symbols $s_j[n]$ and the received signal in mode i, $y_i(t)$, as:

$$y_i(t) = \sum_{n} \sum_{j=1}^{D} s_j[n]q_{ij}(t - nT) + n_i(t) \tag{15}$$

with $n_i(t)$ as the noise in the i-th receiver mode.

4.2. Linear MMSE MIMO Receiver

The most widely used linear MIMO receiver for SDM systems is based on the design of filter $\underline{\underline{Q}}(\omega)$ in Figure 1 to minimize the mean squared error (MSE), which is called a linear MMSE MIMO receiver [28,37,39]. Mathematically, the MSE for the linear MIMO receiver under the MMSE criterion is defined as:

$$\sigma_{\text{MMSE-LE}}^2 = \mathrm{E}\left[\underline{e}^H[n]\underline{e}[n]\right] \tag{16}$$

where

$$\underline{e}[n] = \underline{s}[n] - \underline{\hat{s}}[n] \tag{17}$$

with $\underline{\hat{s}}[n]$ as the output vector of the linear MIMO receiver $\underline{\underline{Q}}(\omega)$.

It is well known that the structure of a linear MMSE MIMO receiver can be divided into a matched filter $\underline{\underline{Q}}^H(\omega)$ operating in continuous time, a sampler operating at the symbol rate, and a discrete-time equalizer of response $\underline{\underline{W}}(\Omega)$, as presented in Figure 2.

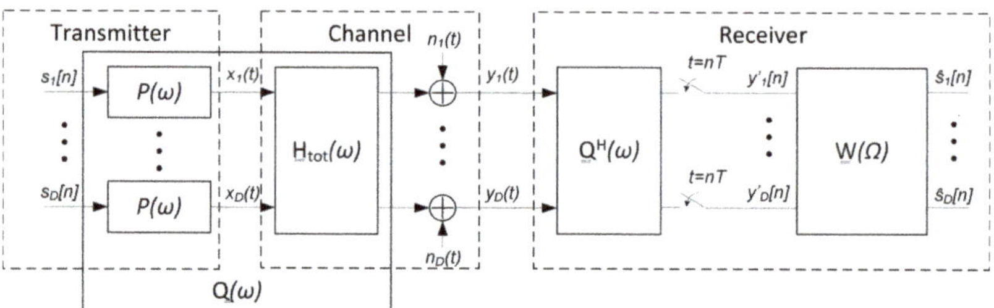

Figure 2. SDM communication system model with linear minimum mean square error (MMSE) MIMO receiver.

We denote as $\underline{y}'[n]$ the vector of samples after the match filter $\underline{\underline{Q}}^H(\omega) = \mathcal{F}\{\underline{\underline{q}}^H(-t)\}$ in the receiver and a symbol rate sampling, being $\underline{\underline{q}}(t) = \mathcal{F}^{-1}\{\underline{\underline{Q}}(\omega)\}$ defined in (14). Now, we define the sampled impulse response at $t = nT$ of the convolution of $\underline{\underline{q}}(t)$ and its matched filter $\underline{\underline{q}}^H(-t)$, which represents the equivalent discrete channel response as:

$$\underline{\underline{g}}[n] = \underline{\underline{q}}(t) * \underline{\underline{q}}^H(-t)\Big|_{t=nT'} \tag{18}$$

and its discrete Fourier transform pair as:

$$\underline{\underline{G}}(\Omega) = \sum_{n=-\infty}^{\infty} \underline{\underline{g}}[n] \cdot e^{-j\Omega n}. \tag{19}$$

Hence, the optimal discrete-time MIMO equalizer $\underline{\underline{W}}_{opt}(\Omega)$ according to the MMSE criterion becomes:

$$\underline{\underline{W}}_{opt}(\Omega) = \left[\underline{\underline{G}}(\Omega) + \underline{\underline{I}} \cdot \left(\frac{N_0}{2}\right)\right]^{-1}, \tag{20}$$

where we are considering a normalized transmission power equally distributed in each of the D modes.

When $\underline{\underline{G}}(\Omega)$ satisfies the Nyquist criterion for MIMO systems $\underline{\underline{G}}(\Omega) = \underline{\underline{I}}$, there is neither ISI nor cross-talk at the matched filter output, further equalization would not be needed, and the optimum linear receiver consists only in the matched filter. However, if such a criterion is not fulfilled, the equalizer $\underline{\underline{W}}(\Omega)$ is essential and some SNDR loss at the output will be unavoidable w.r.t. the ideal case.

4.3. Matched Filter-Based Receiver for SDM

In this subsection we will explore the optical channel requirements to reduce the linear MIMO receiver $\underline{Q}(\omega)$ in Figure 1 to a simple matched filter-based receiver. Furthermore, we will show that the resulting receiver is optimal in the sense that the discrete-time system response of the SDM communication system is the identity matrix, followed by the addition of the AWGN noise.

Let us first write out:

$$\underline{Q}(\omega) = P(\omega) \cdot \underline{\underline{H}}_{tot}(\omega), \tag{21}$$

where $\underline{\underline{H}}_{tot}(\omega)$ and $P(\omega)$ are the Fourier transforms of $\underline{h}_{tot}(t)$ and $p(t)$, respectively and according to what is plotted in Figure 2. It follows that:

$$\underline{Q}(\omega)^H = P^*(\omega) \cdot \underline{\underline{H}}_{tot}(\omega)^H. \tag{22}$$

The signal at each of the D branches $y_i(t)$, defined in (15), is processed before sampling by the continuous-time filter $\underline{Q}^H(\omega)$. The equivalent scheme for this matched filter-based MIMO receiver is shown in Figure 3a. By using the linearity of the system we can rearrange Figure 3a to obtain Figure 3b. Then, elaborating the expression $\underline{Q}(\omega)\underline{Q}^H(\omega)$ we obtain that:

$$\underline{Q}(\omega)\underline{Q}^H(\omega) = P(\omega) \cdot \underline{\underline{H}}_{tot}(\omega) \cdot \underline{\underline{H}}_{tot}^H(\omega) \cdot P^*(\omega) = P(\omega) \cdot \underline{\underline{H}}(\omega) \cdot \underline{\underline{H}}^H(\omega) \cdot P^*(\omega), \tag{23}$$

where we have used that:

$$H_{CD}(\omega) \cdot H_{CD}^*(\omega) = 1. \tag{24}$$

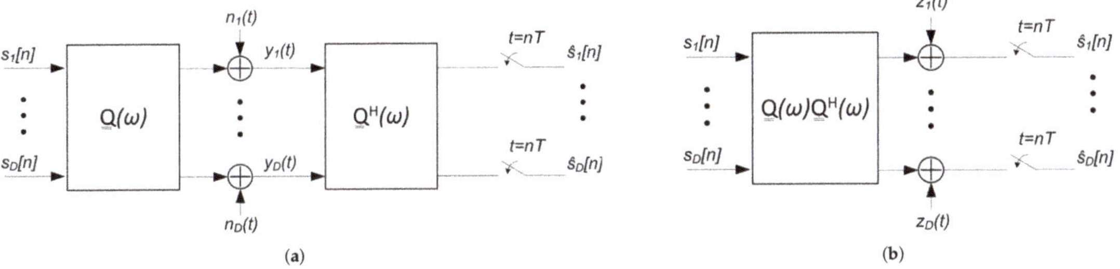

Figure 3. SDM communication system model with matched filter-based receiver (**a**) and its reordered version (**b**).

From (10) we have that:

$$\underline{\underline{H}}(\omega) \cdot \underline{\underline{H}}(\omega)^H =$$

$$\left(\prod_{k=1}^{K_{amp}-1} \underline{\underline{V}}^{(k)} \underline{\underline{\Lambda}}^{(k)}(\omega) \underline{\underline{U}}^{(k)H} \right) \cdot \underline{\underline{V}}^{(K_{amp})} \underline{\underline{\Lambda}}^{(K_{amp})}(\omega) \underline{\underline{U}}^{(K_{amp})H}$$

$$\cdot \underline{\underline{U}}^{(K_{amp})} \underline{\underline{\Lambda}}^{(K_{amp})H}(\omega) \underline{\underline{V}}^{(K_{amp})H}$$

$$\cdot \left(\prod_{k=2}^{K_{amp}} \underline{\underline{U}}^{(K_{amp}-k+1)} \underline{\underline{\Lambda}}^{(K_{amp}-k+1)H}(\omega) \underline{\underline{V}}^{(K_{amp}-k+1)H} \right) \tag{25}$$

$$= \left(\prod_{k=1}^{K_{amp}-1} \underline{\underline{V}}^{(k)} \underline{\underline{\Lambda}}^{(k)}(\omega) \underline{\underline{U}}^{(k)H} \right) \cdot \underline{\underline{V}}^{(K_{amp})} \cdot |\underline{\underline{\Lambda}}^{(K_{amp})}(\omega)|^2 \cdot \underline{\underline{V}}^{(K_{amp})H}$$

$$\cdot \left(\prod_{k=2}^{K_{amp}} \underline{\underline{U}}^{(K_{amp}-k+1)} \underline{\underline{\Lambda}}^{(K_{amp}-k+1)H}(\omega) \underline{\underline{V}}^{(K_{amp}-k+1)H} \right).$$

And the diagonal matrix:

$$|\underline{\underline{\Delta}}^{(K_{amp})}(\omega)|^2 =$$

$$\mathrm{diag}\left(\left[e^{\frac{1}{2}g_1^{(K_{amp})}-j\omega\tau_1^{(K_{amp})}},\ldots,e^{\frac{1}{2}g_D^{(K_{amp})}-j\omega\tau_D^{(K_{amp})}}\right]\right)\cdot$$

$$\mathrm{diag}\left(\left[e^{\frac{1}{2}g_1^{(K_{amp})}+j\omega\tau_1^{(K_{amp})}},\ldots,e^{\frac{1}{2}g_D^{(K_{amp})}+j\omega\tau_D^{(K_{amp})}}\right]\right) \tag{26}$$

$$= \left|\mathrm{diag}\left(\left[e^{\frac{1}{2}g_1^{(K_{amp})}},\ldots,e^{\frac{1}{2}g_D^{(K_{amp})}}\right]\right)\right|^2$$

that does not allow simplifying (25) unless the following holds:

$$|\underline{\underline{\Delta}}^{(K_{amp})}(\omega)|^2 = e^{\left(g^{(K_{amp})}\right)}\cdot\underline{\underline{I}}. \tag{27}$$

This latter condition is equivalent to assuming that:

$$e^{\left(\frac{1}{2}g^{(K_{amp})}\right)} = e^{\left(\frac{1}{2}g_1^{(K_{amp})}\right)} = e^{\left(\frac{1}{2}g_2^{(K_{amp})}\right)} = \cdots = e^{\left(\frac{1}{2}g_D^{(K_{amp})}\right)}, \tag{28}$$

or, in other words, that the MDL is negligible for the K_{amp}-th span. When the condition expressed in (27) is satisfied for all the spans of the system, we can commute the terms in (25), and therefore, we can obtain:

$$\underline{\underline{H}}(\omega)\cdot\underline{\underline{H}}(\omega)^H = \prod_{k=1}^{K_{amp}}|\underline{\underline{\Delta}}^{(k)}(\omega)|^2 = \prod_{k=1}^{K_{amp}}e^{\left(g^{(k)}\right)}\cdot\underline{\underline{I}} = e^{\left(\sum_{k=1}^{K_{amp}}g^{(k)}\right)}\cdot\underline{\underline{I}}. \tag{29}$$

Revisiting (23), and plugging in (29) under the assumption of a negligible MDL in the optical channel, we can write:

$$\underline{\underline{G}}(\omega) = \underline{\underline{Q}}(\omega)\cdot\underline{\underline{Q}}^H(\omega) = |P(\omega)|^2\cdot e^{\left(\sum_{k=1}^{K_{amp}}g^{(k)}\right)}\cdot\underline{\underline{I}}. \tag{30}$$

Therefore, without loss of generality, $e^{\left(\sum_{k=1}^{K_{amp}}g^{(k)}\right)} = 1$ can be assumed and, after sampling at the symbol rate, (30) can be written as:

$$\underline{\underline{G}}(\Omega) = \frac{1}{T}\cdot\sum_{l=-\infty}^{\infty}\left|P\left(\frac{\Omega+2\pi l}{T}\right)\right|^2\cdot\underline{\underline{I}} = \underline{\underline{I}}, \tag{31}$$

where we have used that $p(t)$ defined in Equation (12) is a square-root raised cosine pulse satisfying the Nyquist criterion.

Regarding the filtered noise waveforms $z_1(t)$ to $z_D(t)$ in Figure 4a, they have an autocorrelation function matrix $\underline{\underline{R}}_{zz}(t)$, whose Fourier transform pair $\underline{\underline{S}}_z(\omega)$ can be expressed as:

$$\underline{\underline{S}}_z(\omega) = \underline{\underline{Q}}(\omega)\cdot\underline{\underline{S}}_n(\omega)\cdot\underline{\underline{Q}}^H(\omega), \tag{32}$$

where $\underline{\underline{S}}_n(\omega)$ is the Fourier transform of the autocorrelation function matrix $\underline{\underline{R}}_{nn}(t) = \underline{\underline{I}}\cdot\frac{N_0}{2}\cdot\delta(t)$ of the received noise vector $\underline{n}(t) = [n_1(t), n_2(t),\ldots,n_D(t)]^T$. We remind that the noise components of the noise vector $\underline{n}(t)$ were assumed uncorrelated with identical power in each mode equal to $N_0/2$. Using (30) and (32) leads to:

$$\underline{\underline{S}}_z(\omega) = \frac{N_0}{2}\cdot\underline{\underline{Q}}(\omega)\cdot\underline{\underline{Q}}^H(\omega) = \frac{N_0}{2}\cdot|P(\omega)|^2\cdot e^{\left(\sum_{k=1}^{K_{amp}}g^{(k)}\right)}\cdot\underline{\underline{I}}. \tag{33}$$

Using the previous assumptions about gains $g^{(k)}$ and $P(\omega)$ made before, we obtain that the sampled noise vector $\underline{z}[n] = [z_1[n], \ldots, z_D[n]]^T$ has an autocorrelation matrix function $\underline{\underline{R}}_{zz}[n] = \underline{\underline{I}} \cdot \frac{N_0}{2} \cdot \delta[n]$.

Therefore, we can conclude that a $D \times D$ MIMO coherent optical communication system using a continuous-time matched filter as a receiver completely eliminates channel ISI and crosstalk when the MDL in the channel is negligible. Moreover, the equivalent discrete-time system model reduces to D discrete parallel AWGN channels as shown in Figure 4b. Hence, there would be no loss of performance w.r.t. the AWGN channel without distortion.

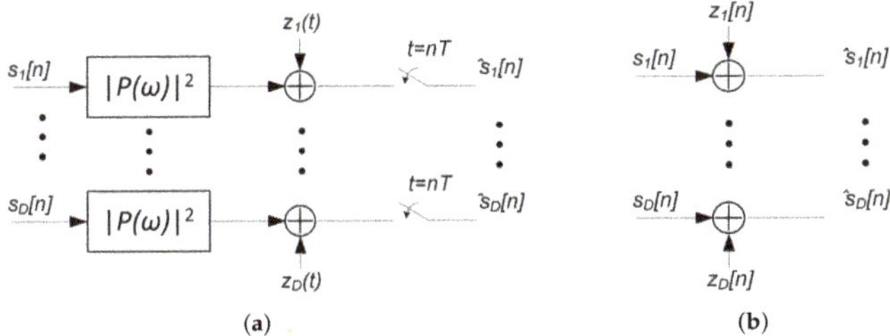

Figure 4. MIMO coherent optical communication system model with matched filter-based receiver in the absence of mode-dependent loss (MDL) (**a**) and its equivalent discrete-time system model (**b**).

5. Numerical Simulation of Linear MIMO FSE Receiver for MDL-Impaired Optical Channel

In this section, we assess the performance of the ideal MMSE linear receiver when MDL is present. Specifically, we study the SNDR degradation at the receiver output w.r.t. the case when the MDL is negligible. To carry out this study, we will use a FSE-based receiver, as shown in Figure 5, which is the most common implementation of the ideal linear filter in discrete-time systems (see $\underline{\underline{Q}}(\omega)$ in Figure 1). Note that this scheme is only valid for integer oversampling rates r_{ov}.

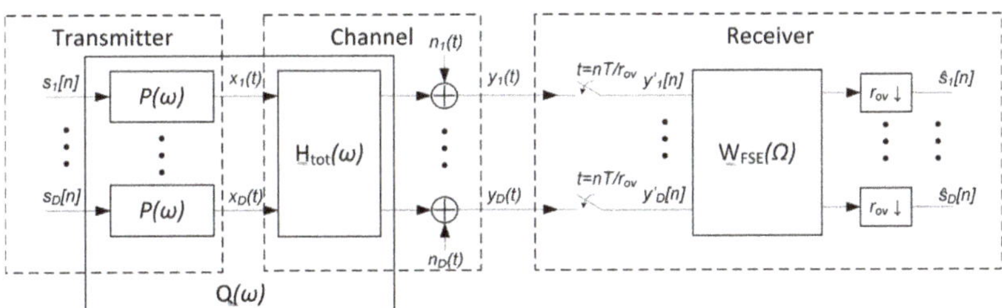

Figure 5. SDM communication system model with linear fractionally-spaced equalizer (FSE) MIMO receiver and integer oversampling rate r_{ov}.

The FSE oversampling rate r_{ov} has been set to two [1,12] and the discrete-time equalizer, with a $\underline{\underline{W}}_{FSE}(\Omega)$ response, has been designed with a number of taps N_{taps} large enough so that any further increase does not lead to a significantly better SNDR at the receiver output. The decimated output of $\underline{\underline{W}}_{FSE}(\Omega)$, by a r_{ov} factor, are the estimated symbol $\underline{\hat{s}}[n]$.

We define the receiver performance metric for each mode i, $L(i)$, as the difference in dB between the output SNR of an ISI and crosstalk-free system with D parallel AWGN channels (see Figure 4b), and the output SNDR of the FSE-based receiver, denoted as $\underline{SNDR}_{out}$.

5.1. Channel Model

We decide to carry out the numerical simulations of the $\underline{H}(\omega)$ channel model described in (6) in the time domain, so that the relative delays of the different modes can be easily described as a time shift between them. The different mode amplitudes can also be handled simply by a diagonal matrix. The chromatic dispersion, which have a SISO frequency response $H_{CD}(\omega)$ that does not depend on the mode $i \in \{0, \ldots, D\}$, is represented as [7]:

$$H_{CD}(\omega) = e^{-j\beta \frac{\omega^2}{2}}, \tag{34}$$

where $\beta = \bar{\beta}_2 \ell_{tot}$, and $\ell_{tot} = K_{amp} \ell_{span}$ when all spans are considered of equal length.

The MDL effect is modeled with an amplification factor for each mode and each optical amplifier (located at the end of each span). These factors are considered time-invariant for a given channel realization in the form of a vector for the k-th span $\underline{g}^{(k)} = [g_1^{(k)}, g_2^{(k)}, \ldots, g_D^{(k)}]$, where $g_i^{(k)}$ for $i \in \{0, \ldots, D\}$ is expressed in dB and taken from a Gaussian distribution with zero mean and standard deviation (STD) σ_g. The sum of all factors $\sum_{i=1}^{D} g_i^{(k)}$ is set to 0 for normalization purposes. Hence, the amplitudes matrix of the k-th span, frequency independent, is given by:

$$\underline{\underline{A}}^{(k)} = \mathrm{diag}\left(\left[e^{\left(\frac{1}{2}g_1^{(k)}\right)}, \ldots, e^{\left(\frac{1}{2}g_D^{(k)}\right)}\right]\right). \tag{35}$$

Alternatively, for each span k of the communication link we have the delays matrix:

$$\underline{\underline{\Lambda}}^{(k)}(\omega) = \underline{\underline{A}}^{(k)} \cdot \mathrm{diag}\left(\left[e^{-j\omega\tau_1^{(k)}}, \ldots, e^{-j\omega\tau_D^{(k)}}\right]\right), \tag{36}$$

being $\underline{\tau}^{(k)} = [\tau_1^{(k)}, \tau_2^{(k)}, \ldots, \tau_D^{(k)}]$ the vector that models the MD with group delays for each mode of the k-th span.

To obtain the delays, we generate the first $D/2$ values of $\underline{\tau}^{(k)}$ from a Gaussian distribution with STD σ_{gd}, and the second $D/2$ values are taken as the opposite of these, which satisfies that $\sum_{i=1}^{D} \tau_i^{(k)} = 0$, since we consider that the system uses polarization multiplexing as part of the SDM [40].

The time-domain impulse response for each of the spans k is calculated by applying the inverse Fourier transform to (7) and can be expressed as:

$$\underline{h}^{(k)}(t) = \underline{\underline{V}}^{(k)} \underline{\underline{A}}^{(k)} \underline{\underline{d}}^{(k)}(t) \underline{\underline{U}}^{(k)}, \tag{37}$$

where

$$\underline{\underline{d}}^{(k)}(t) = \mathrm{diag}\left(\left[\delta(t - \tau_1^{(k)}), \ldots, \delta(t - \tau_D^{(k)})\right]\right), \tag{38}$$

and we have used that the matrices $\underline{\underline{A}}^{(k)}$, $\underline{\underline{U}}^{(k)}$, and $\underline{\underline{V}}^{(k)}$ are constant.

Equation (37) describes that incoming signal at the kth span is multiplied by the unitary matrix $\underline{\underline{U}}^{(k)}$, then each modal impulse response is delayed by $\tau_i^{(k)}$, the amplification factor is set by the diagonal matrix $\underline{\underline{A}}^{(k)}$ and the mode-mixing unitary matrix $\underline{\underline{V}}^{(k)}$ is applied. Finally, the impulse response of the complete channel is given by:

$$\underline{h}_{tot}(t) = \underline{h}^{(K_{amp})}(t) * \underline{h}^{(K_{amp}-1)}(t) * \cdots * \underline{h}^{(1)}(t) * h_{CD}(t), \tag{39}$$

where $h_{CD}(t) = \mathcal{F}^{-1}\{H_{CD}(\omega)\}$.

Note that, due to the random nature of $\underline{g}^{(k)}$ and $\underline{\tau}^{(k)}$ in each k span, we can generate an arbitrary number N_{ch} of channel realizations of $\underline{\underline{H}}_{tot}(\omega) = \mathcal{F}\{\underline{h}_{tot}(t)\}$ for a given value of σ_g and σ_{gd}.

Since we consider all the modes to be strongly coupled, the $\underline{\underline{U}}^{(k)}$ and $\underline{\underline{V}}^{(k)}$ matrices of each span k are modeled as unitary Gaussian random matrices obtained from a QR factorization of a complex random matrix whose elements have a zero mean and STD equal to 1. The two orthogonal matrices after QR factorization of two independent realizations of the random matrix are used as $\underline{\underline{U}}^{(k)}$ and $\underline{\underline{V}}^{(k)}$, respectively.

We consider a total number of $K_{amp} = 100$ spans, each $\ell_{span} = 50$ km long. For the fiber parameters, we used the multi-core fiber data reported in [41], considering the number of modes $D = 6$ and the central wavelength $\lambda_c = 1469$ nm. The selection of this multi-core fiber allow us to compare the results of this work with those presented in [28], and to obtain the fiber parameters needed for the numerical simulations from [41].

We take 2% as the underestimation dispersion factor that is applied to the dispersion coefficient D_{CD} to obtain the residual CD experienced by the receiver. For the gain STD σ_g, we considered several values in the range of the systems referenced in [27]. For the numerical simulations, we compute a total of $N_{ch} = 10{,}000$ realizations of the channel frequency response $\underline{\underline{H}}_{tot}(\omega)$ defined in (5).

5.2. Transmitter and Linear MIMO FSE Receiver Parameters

As described in Section 4, the transmitter uses a generalized PAM modulation and an square-root raised cosine for pulse shaping with a roll-off factor equal to α. We will show results of the numerical simulation for several values of α. The symbol period T has been set for a symbol rate $R_s = 64$ GBaud. The FSE-based MIMO receiver has an oversampling factor $r_{ov} = 2$, and a number of equalizer taps $N_{taps} = 1000$ has been selected to ensure that it does not limit the receiver performance for the considered channel MDL.

5.3. Signal-to-Noise at the Input of the Receiver

The signal-to-noise ratio at the input of each mode $\left(\frac{S}{N}\right)_{in}(i)$ for $i \in \{0,\dots,D\}$ is defined as:

$$\left(\frac{S}{N}\right)_{in}(i) = \frac{P_{in}(i)}{N_0/2}. \tag{40}$$

The signal-to-noise at the input of the receiver $\overline{SNR}_{in}$ in dB can be written as:

$$\overline{SNR}_{in} = 10 \cdot \log_{10}\left(\frac{\frac{1}{D} \cdot \sum_{i=1}^{D} P_{in}(i)}{N_0/2}\right) = 10 \cdot \log_{10}\left(\frac{1}{D} \cdot \sum_{i=1}^{D} \left(\frac{S}{N}\right)_{in}(i)\right), \tag{41}$$

and it is taken from the set of values in Table 1. $P_{in}(i)$ is the receiver input power in the mode i for the current channel realization.

5.4. Performance Loss Metric for FSE-Based MIMO Receiver

We define the performance loss metric (in dB) of the FSE-based MIMO receiver in MDL-impaired channels for certain mode i as:

$$L(i) = SNR_{in}(i) - SNDR_{out}(i) \tag{42}$$

where $SNR_{(in)}(i) = 10 \cdot \log_{10}\left(\frac{S}{N}\right)_{in}(i)$, and $\underline{SNDR}_{out} = [SNDR_{out}(1), SNDR_{out}(2),\dots,$ $SNDR_{out}(D)]^T$ is calculated as defined in [42] and Equation (28) in [43] for MIMO implementation of the FSE. Given a set of system model parameters, the numerical simulation will generate a total of $D \cdot N_{ch}$ values of $L(i)$. The average loss AL is calculated for each channel realization of among the available N_{ch} as:

$$AL = \frac{1}{D} \sum_{i=1}^{D} L(i). \tag{43}$$

Two FSE-based MIMO receiver performance metrics can be extracted from the $D \cdot N_{ch}$ calculated values of $L(i)$:

- ML_{95} is defined as the 95th percentile of the $L(i)$ distribution obtained for any optical channel realization and mode;

- AML_{95} is defined as the 95th percentile of the AL distribution obtained for any optical channel realization.

The values for the parameters used in the simulation are summarized in Table 1.

Table 1. Simulation parameters.

Parameter	Symbol	Value and Reference
Span length	ℓ_{span}	50 km
Number of spans	K_{amp}	100
Number of spatial and polarization modes	D	6
Center wavelength	λ_c	1469 nm [41]
Modal dispersion	$\sigma_\tau / \sqrt{\ell_{span}}$	3.1 ps/$\sqrt{\text{km}}$ [41]
Dispersion coefficient	$D_{CD} = -\frac{2\pi c}{\lambda_c^2}\bar{\beta}_2$	20.1 ps/(nm·km) [41]
Underestimation dispersion factor	U_{CD}	2% [9]
Amplifier gain STD	σ_g	0, 0.1, 0.2, 0.3, 0.4, 0.5, 0.6 dB
Symbol rate	$R_s = 1/T_s$	64 GBaud
Oversampling factor	r_{ov}	2
Roll off factor	α	0.1, 0.5, 0.7, 0.9
Number of channel realizations	N_{ch}	10000
Signal to noise ratio at the receiver input	$\overline{SNR}_{in}$	60 30 15 10 6.2 5 dB [27]
Number of taps	N_{taps}	1000

5.5. Numerical Simulation Results

The first result is focused on the impact of PAM pulses roll-off factor α and MDL level, represented by σ_g, on the SDM optical system performance. Figure 6 shows that for systems with transmitters using a higher α, the degradation is a bit lower. The effect is higher with increasing σ_g for systems working at a high regime of $\overline{SNR}_{in}$, as seen in Figure 6b.

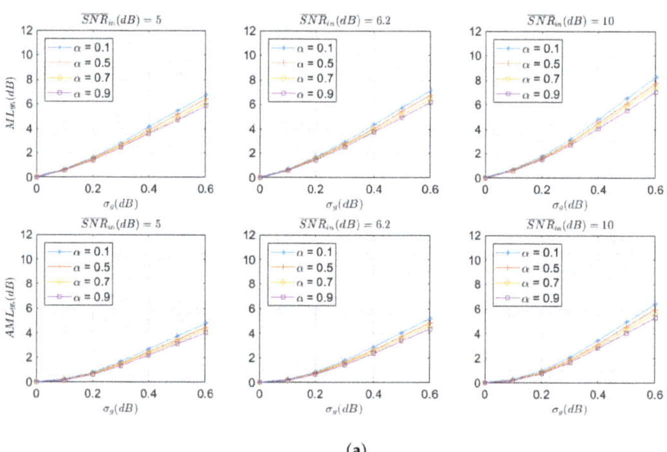

(a)

Figure 6. *Cont.*

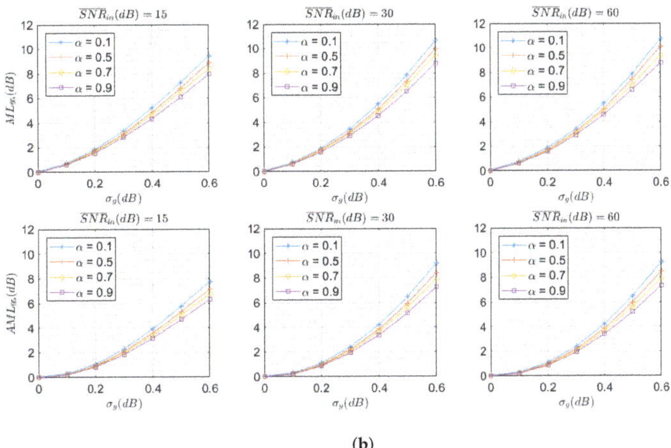

(b)

Figure 6. ML_{95} (up) and AML_{95} (down) as defined in Section 5.4 for $\overline{SNR}_{in}$ = 5, 6.2, and 10 dB (**a**) and $\overline{SNR}_{in}$ = 15, 30, and 60 dB (**b**) for different values of the transmitter roll-off factor α. Note that $\sigma_g = 0$ corresponds to a channel without MDL.

In practical systems, the allowable loss of $SNDR_{out}$ in a channel with elements introducing MDL w.r.t. an ideal channel without MDL is around 1–2 dB. We can observe that, assuming a maximum degradation of 2 dB in the system with a 95% confidence, σ_g values of the amplifiers should not exceed 0.2 dB. These results are in agreement with the capacity limits of a MIMO MMSE receiver and MDL channel calculated in [27].

A second result is presented in Figure 7, where the probability distribution of AL and $\overline{SNR}_{in}$ estimated from the analysis of all channel realizations is plotted. We are comparing different levels of SNR regimes, with $\overline{SNR}_{in}$ = 5, 6.2, and 10 dB (Figure 7a) and $\overline{SNR}_{in}$ = 15, 30, and 60 dB (Figure 7b), for a roll-off factor of $\alpha = 0.9$. The upper and lower limits of the blue boxes represent the 25th and 75th percentiles respectively. The red line inside the box indicates the median of the metric. In case the distribution of values was Gaussian, the whisker bounds correspond to 2.7 times the STD of the metric or, in other words, the number of values between the upper and lower bounds of the whiskers contains 99.3% of the values. Values outside these limits are considered outliers and are individually represented by red crosses.

We make the following observations from Figure 7:

- The distribution of AL is not Gaussian, as we can observe by comparing the difference between upper and lower outliers for higher σ_g values and their asymmetry;
- There are no negative values of AL, since the FSE MIMO receiver cannot improve on average the $\overline{SNR}_{in}$. However, by taking all values of $L(i)$ for any received mode i, we can find that, for certain channels and modes, the FSE MIMO receiver can locally improve the $SNR_{in}(i)$ of a particular mode i, but always at the cost of another mode of the receiver;
- The performance degradation depends on the $\overline{SNR}_{in}$. In a low $\overline{SNR}_{in}$ regime (5 dB, Figure 7a), the degradation is measured lower in absolute values when compared to the high $\overline{SNR}_{in}$ regime (60 dB, Figure 7b);
- The STD of the performance degradation also depends on the $\overline{SNR}_{in}$. In the low $\overline{SNR}_{in}$ regime (5 dB, Figure 7a), the STD of the degradation is lower when compared to the high $\overline{SNR}_{in}$ regime (60 dB, Figure 7b);
- The performance degradation measured as AL is milder than measured as $L(i)$ when more than 95% coverage of the channels is considered. Note that the difference is negligible when median values are taken into account.

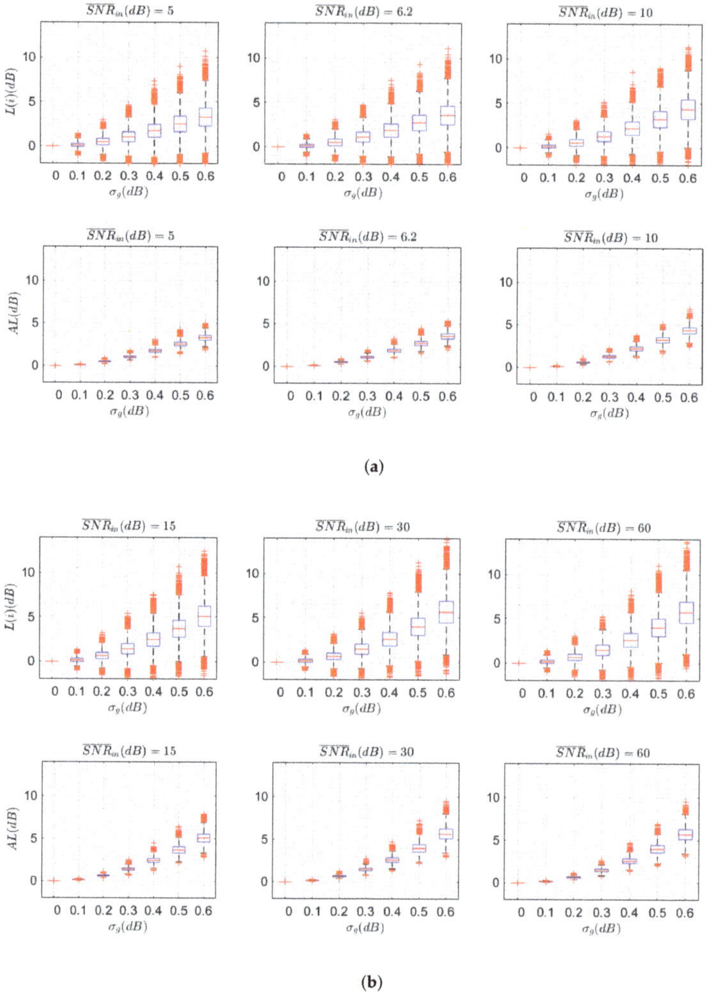

Figure 7. Probability distribution for AL as defined in (43) (up) and $L(i)$ as defined in (42) (down) for $\overline{SNR}_{in}$ = 5, 6.2, and 10 dB (**a**) and $\overline{SNR}_{in}$ = 15, 30, and 60 dB (**b**). $\alpha = 0.9$ for all graphs. Note that $\sigma_g = 0$ corresponds to a channel without MDL.

6. Conclusions

This work explored long-haul fiber-optic SDM coherent systems with PAM raised-cosine pulses. We investigated the overall system performance under different configurations of the optical channel. For that purpose, we modeled this channel with a MIMO multi-span structure that included the several dispersion terms and modal losses factors.

It was demonstrated that the linear MMSE MIMO receiver completely eliminated ISI and crosstalk when the number of taps was sufficiently high and the optical channel was free of MDL. Moreover, the generic structure of the linear MMSE MIMO receiver could be simplified to a continuous-time matched filter and still retained the same properties. This observation paves the way for analog receivers that simply implement the matched filter of the optical channel, eliminating all channel impairments when MDL is negligible.

We also defined performance metrics to assess the losses of a linear MIMO receiver implemented using a MIMO FSE with an oversampling factor of two for an optical channel that exhibited significant MDL. We have shown that such loss depends on the transmitter PAM pulses roll-off factor and the SNR level at the receiver input. We also determined that the performance degradation could be limited by processing the D output modes together by averaging the D output SNDR values at the receiver.

This fact opens a way to exploit this loss compensation between modes at the receiver output. The design of specific forward error correction codes taking into account this aspect could improve the final performance of the system in terms of bit error rate. Constructing the message to be encoded, including bits or signals belonging to all modes, could improve system performance w.r.t. constructing messages with bits or signals from only one mode.

Author Contributions: Conceptualization, methodology, validation, formal analysis, resources, investigation, data curation, writing—review and editing, writing—original draft preparation, and visualization, L.M.T., F.J.C. and L.D.; software, L.M.T. and L.D.; supervision, project administration, and funding acquisition F.J.C. and L.D. All authors have read and agreed to the published version of the manuscript.

Funding: This research was supported in part by Junta de Andalucía, under PAIDI program, TIC-102 research group, and by the Spanish Ministry of Ciencia e Innovación, under Project PID2019-109842RB-I00.

Institutional Review Board Statement: Not applicable.

Data Availability Statement: Not applicable.

Conflicts of Interest: The authors declare no conflict of interest.

References

1. Shibahara, K.; Mizuno, T.; Lee, D.; Miyamoto, Y. Advanced MIMO Signal Processing Techniques Enabling Long-Haul Dense SDM Transmissions. *J. Light. Technol.* **2017**, *36*, 336–348. [CrossRef]
2. Arik, S.O.; Kahn, J.M.; Ho, K.P. MIMO Signal Processing for Mode-Division Multiplexing: An overview of channel models and signal processing architectures. *IEEE Signal Process. Mag.* **2014**, *31*, 25–34. [CrossRef]
3. Evangelides, S.G.; Mollenauer, L.F.; Gordon, J.P.; Bergano, N.S. Polarization multiplexing with solitons. *J. Light. Technol.* **1992**, *10*, 28–35. [CrossRef]
4. Hsu, R.C.; Tarighat, A.; Shah, A.; Sayed, A.H.; Jalali, B. Capacity enhancement in coherent optical MIMO (COMIMO) multimode fiber links. *IEEE Commun. Lett.* **2006**, *10*, 195–197. [CrossRef]
5. Tarighat, A.; Hsu, R.C.; Shah, A.; Sayed, A.H.; Jalali, B. Fundamentals and challenges of optical multiple-input multiple-output multimode fiber links [Topics in Optical Communications]. *IEEE Commun. Mag.* **2007**, *45*, 57–63. [CrossRef]
6. Betti, S.; Curti, F.; De Marchis, G.; Iannone, E. A novel multilevel coherent optical system: 4-quadrature signaling. *J. Light. Technol.* **1991**, *9*, 514–523. [CrossRef]
7. Ip, E.; Kahn, J.M. Digital Equalization of Chromatic Dispersion and Polarization Mode Dispersion. *J. Light. Technol.* **2007**, *25*, 2033–2043. [CrossRef]
8. Forestieri, E.; Prati, G. Exact Analytical Evaluation of Second-Order PMD Impact on the Outage Probability for a Compensated System. *J. Light. Technol.* **2004**, *22*, 988–996. [CrossRef]
9. Ip, E.; Lau, A.P.T.; Barros, D.; Kahn, J.M. Coherent detection in optical fiber systems. *Opt. Express* **2008**, *16*, 753–791. [CrossRef]
10. Savory, S.J. Digital filters for coherent optical receivers. *Opt. Express* **2008**, *16*, 804–817. [CrossRef]
11. Savory, S.J. Digital Coherent Optical Receivers: Algorithms and Subsystems. *IEEE J. Sel. Top. Quantum Electron.* **2010**, *16*, 1164–1179. [CrossRef]
12. Spinnler, B. Equalizer Design and Complexity for Digital Coherent Receivers. *IEEE J. Sel. Top. Quantum Electron.* **2010**, *16*, 1180–1192. [CrossRef]
13. Ryf, R.; Randel, S.; Gnauck, A.H.; Bolle, C.; Sierra, A.; Mumtaz, S.; Esmaeelpour, M.; Burrows, E.C.; Essiambre, R.J.; Winzer, P.J.; et al. Mode-Division Multiplexing Over 96 km of Few-Mode Fiber Using Coherent 6×6 MIMO Processing. *J. Light. Technol.* **2011**, *30*, 521–531. [CrossRef]
14. Sillard, P.; Molin, D.; Bigot-Astruc, M.; Amezcua-Correa, A.; de Jongh, K.; Achten, F. 50 um Multimode Fibers for Mode Division Multiplexing. *J. Light. Technol.* **2016**, *34*, 1672–1677. [CrossRef]
15. Shibahara, K.; Lee, D.; Kobayashi, T.; Mizuno, T.; Takara, H.; Sano, A.; Kawakami, H.; Miyamoto, Y.; Ono, H.; Oguma, M.; et al. Dense SDM (12-Core x 3-Mode) Transmission Over 527 km with 33.2-ns Mode-Dispersion Employing Low-Complexity Parallel MIMO Frequency-Domain Equalization. *J. Light. Technol.* **2016**, *34*, 196–204. [CrossRef]
16. Rademacher, G.; Luis, R.S.; Puttnam, B.J.; Eriksson, T.A.; Ryf, R.; Agrell, E.; Maruyama, R.; Aikawa, K.; Awaji, Y.; Furukawa, H.; et al. High Capacity Transmission with Few-Mode Fibers. *J. Light. Technol.* **2018**, *37*, 425–432. [CrossRef]

17. van Weerdenburg, J.; Ryf, R.; Alvarado-Zacarias, J.C.; Alvarez-Aguirre, R.A.; Fontaine, N.K.; Chen, H.; Amezcua-Correa, R.; Sun, Y.; Gruner-Nielsen, L.; Jensen, R.V.; et al. 138-Tb/s Mode- and Wavelength-Multiplexed Transmission Over Six-Mode Graded-Index Fiber. *J. Light. Technol.* **2018**, *36*, 1369–1374. [CrossRef]

18. Ho, K.P.; Kahn, J.M. Mode-dependent loss and gain: Statistics and effect on mode-division multiplexing. *Opt. Express* **2011**, *19*, 16612–16635. [CrossRef] [PubMed]

19. Antonelli, C.; Mecozzi, A.; Shtaif, M.; Winzer, P.J. Stokes-space analysis of modal dispersion in fibers with multiple mode transmission. *Opt. Express* **2012**, *20*, 11718–11733. [CrossRef]

20. Ho, K.P.; Kahn, J.M. Mode Coupling and its Impact on Spatially Multiplexed Systems. In *Optical Fiber Telecommunications*; Elsevier: Amsterdam, The Netherlands, 2013; pp. 491–568. [CrossRef]

21. Arik, S.O.; Askarov, D.; Kahn, J.M. Adaptive Frequency-Domain Equalization in Mode-Division Multiplexing Systems. *J. Light. Technol.* **2014**, *10*, 1841–1852. [CrossRef]

22. Inan, B.; Spinnler, B.; Ferreira, F.; van den Borne, D.; Lobato, A.; Adhikari, S.; Sleiffer, V.A.; Kuschnerov, M.; Hanik, N.; Jansen, S.L. DSP complexity of mode-division multiplexed receivers. *Opt. Express* **2012**, *20*, 10859–10869. [CrossRef] [PubMed]

23. Arık, S.Ö.; Askarov, D.; Kahn, J.M. Effect of mode coupling on signal processing complexity in mode-division multiplexing. *J. Light. Technol.* **2012**, *31*, 423–431. [CrossRef]

24. Randel, S.; Winzer, P.J.; Montoliu, M.; Ryf, R. Complexity analysis of adaptive frequency-domain equalization for MIMO-SDM transmission. In Proceedings of the 39th European Conference and Exhibition on Optical Communication (ECOC 2013), London, UK, 22–26 September 2013; pp. 1–3.

25. Guiomar, F.P.; Pinto, A.N. Simplified Volterra Series Nonlinear Equalizer for Polarization-Multiplexed Coherent Optical Systems. *J. Light. Technol.* **2013**, *31*, 3879–3891. [CrossRef]

26. Miller, D.A.B. Self-configuring universal linear optical component [Invited]. *Photonics Res.* **2013**, *1*, 1. PRJ.1.000001. [CrossRef]

27. Mello, D.A.A.; Srinivas, H.; Choutagunta, K.; Kahn, J.M. Impact of Polarization- and Mode-Dependent Gain on the Capacity of Ultra-Long-Haul Systems. *J. Light. Technol.* **2019**, *38*, 303–318. [CrossRef]

28. Ospina, R.S.B.; van den Hout, M.; Alvarado-Zacarias, J.C.; Antonio-Lopez, J.E.; Bigot-Astruc, M.; Correa, A.A.; Sillard, P.; Amezcua-Correa, R.; Okonkwo, C.; Mello, D.A.A. Mode-Dependent Loss and Gain Estimation in SDM Transmission Based on MMSE Equalizers. *J. Light. Technol.* **2020**, *39*, 1968–1975. [CrossRef]

29. Ip, E.; Kahn, J. Fiber Impairment Compensation Using Coherent Detection and Digital Signal Processing. *J. Light. Technol.* **2009**, *28*, 502–519. [CrossRef]

30. Antonelli, C.; Mecozzi, A.; Shtaif, M.; Winzer, P.J. Modeling and performance metrics of MIMO-SDM systems with different amplification schemes in the presence of mode-dependent loss. *Opt. Express* **2015**, *23*, 2203. [CrossRef]

31. Yang, S.; Hanzo, L. Fifty years of MIMO detection: The road to large-scale MIMOs. *IEEE Commun. Surv. Tutor.* **2015**, *17*, 1941–1988. [CrossRef]

32. Yang, Z.; Yu, W.; Peng, G.; Liu, Y.; Zhang, L. Recent Progress on Novel DSP Techniques for Mode Division Multiplexing Systems: A Review. *Appl. Sci.* **2021**, *11*, 1363. [CrossRef]

33. Ho, K.P.; Kahn, J.M. Linear Propagation Effects in Mode-Division Multiplexing Systems. *J. Light. Technol.* **2013**, *32*, 614–628. [CrossRef]

34. Choutagunta, K.; Arik, S.O.; Ho, K.P.; Kahn, J.M. Characterizing Mode-Dependent Loss and Gain in Multimode Components. *J. Light. Technol.* **2018**, *36*, 3815–3823. [CrossRef]

35. Choutagunta, K.; Roberts, I.; Kahn, J.M. Efficient Quantification and Simulation of Modal Dynamics in Multimode Fiber Links. *J. Light. Technol.* **2018**, *37*, 1813–1825. [CrossRef]

36. Biglieri, E.; Calderbank, R.; Constantinides, A.; Goldsmith, A.; Paulraj, A.; Poor, H.V. *MIMO Wireless Communication*; Cambridge University Press: Cambridge, UK, 2007.

37. Joham, M.; Utschick, W.; Nossek, J. Linear transmit processing in MIMO communications systems. *IEEE Trans. Signal Process.* **2005**, *53*, 2700–2712. [CrossRef]

38. Cioffi, J.M. Equalization. In *Advanced Digital Communications*; Stanford University Course Notes: Stanford, CA, USA, 2020; pp. 162–394.

39. Li, P.; Paul, D.; Narasimhan, R.; Cioffi, J. On the distribution of SINR for the MMSE MIMO receiver and performance analysis. *IEEE Trans. Inf. Theory* **2005**, *52*, 271–286.

40. Murshid, S.; Grossman, B.; Narakorn, P. Spatial domain multiplexing: A new dimension in fiber optic multiplexing. *Opt. Laser Technol.* **2008**, *40*, 1030–1036. [CrossRef]

41. Hayashi, T.; Tamura, Y.; Hasegawa, T.; Taru, T. Record-Low Spatial Mode Dispersion and Ultra-Low Loss Coupled Multi-Core Fiber for Ultra-Long-Haul Transmission. *J. Light. Technol.* **2017**, *35*, 450–457. [CrossRef]

42. Vandendorpe, L. Fractionally spaced linear and DF MIMO equalizers for multitone systems without guard time. In *Annales des Télécommunications*; Springer: Berlin/Heidelberg, Germany, 1997; Volume 52, pp. 21–30.

43. Vandendorpe, L.; Cuvelier, L.; Deryck, F.; Louveaux, J.; van de Wiel, O. Fractionally spaced linear and decision-feedback detectors for transmultiplexers. *IEEE Trans. Signal Process.* **1998**, *46*, 996–1011. [CrossRef]

Low-Complexity Robust Adaptive Beamforming Based on INCM Reconstruction via Subspace Projection

Yanliang Duan [1], Xinhua Yu [1], Lirong Mei [2,3] and Weiping Cao [1,2,3,*]

[1] Guangxi Key Laboratory of Wireless Wideband Communication & Signal Processing, Guilin University of Electronic Technology at Guilin, Guilin 541004, China; duanyanliangdyl@163.com (Y.D.); yusilian@126.com (X.Y.)

[2] The 54th Research Institute of China Electronics Technology Group Corporation, Shijiazhuang 050081, China; xymeilirong@163.com

[3] Science and Technology on Information Transmission and Dissemination in Communication Networks Laboratory, Shijiazhuang 050081, China

* Correspondence: weipingc@guet.edu.cn

Abstract: Adaptive beamforming is sensitive to steering vector (SV) and covariance matrix mismatches, especially when the signal of interest (SOI) component exists in the training sequence. In this paper, we present a low-complexity robust adaptive beamforming (RAB) method based on an interference–noise covariance matrix (INCM) reconstruction and SOI SV estimation. First, the proposed method employs the minimum mean square error criterion to construct the blocking matrix. Then, the projection matrix is obtained by projecting the blocking matrix onto the signal subspace of the sample covariance matrix (SCM). The INCM is reconstructed by replacing part of the eigenvector columns of the SCM with the corresponding eigenvectors of the projection matrix. On the other hand, the SOI SV is estimated via the iterative mismatch approximation method. The proposed method only needs to know the priori-knowledge of the array geometry and angular region where the SOI is located. The simulation results showed that the proposed method can deal with multiple types of mismatches, while taking into account both low complexity and high robustness.

Keywords: robust adaptive beamforming; orthogonality; blocking matrix; interference-plus-noise covariance matrix reconstruction

Citation: Duan, Y.; Yu, X.; Mei, L.; Cao, W. Low-Complexity Robust Adaptive Beamforming Based on INCM Reconstruction via Subspace Projection. *Sensors* **2021**, *21*, 7783. https://doi.org/10.3390/s21237783

Academic Editor: Jiangbing Du

Received: 5 October 2021
Accepted: 19 November 2021
Published: 23 November 2021

Publisher's Note: MDPI stays neutral with regard to jurisdictional claims in published maps and institutional affiliations.

1. Introduction

Adaptive beamforming adjusts the weight vector according to the application environment to enhance the SOI by suppressing interference and noise, and it has been widely used in radar, sonar, microphone array speech processing, wireless communication, radio astronomy, and other areas [1–3]. Generally, the standard Capon beamformer (SCB) obtains the maximum-array-output-signal-to-interference-plus-noise ratio (SINR) if the covariance matrix and SOI SV are accurately known [4]. However, severe performance degradation may occur in the presence of SV and INCM mismatches due to the fact of array calibration errors, finite snapshots, and other factors, especially when the SOI component is presented in the INCM [5,6]. Therefore, various RAB algorithms have been proposed to ensure the robustness of beamformers over the past years. In general, these RAB methods can be classified into the following types [6,7]: diagonal loading (DL) technique, eigenspace-based (ESB) technique, uncertain-set based technique, and covariance matrix reconstruction-based technique.

DL is one of the most classical RAB methods for improving the robustness of a beamformer, which is derived by imposing a quadratic constraint either on the norm of the weight vector or on its SOI SV [8,9]. However, its major challenge is that it is difficult to choose the optimal DL level in different scenarios. To overcome this drawback, parameter-free methods in [4,10–13] can automatically compute the DL level without specifying any additional user parameters. Regrettably, these methods fail to provide satisfactory performance in high-input signal noise rates (SNRs).

The ESB technique is another type of traditional RAB method that is performed by projecting nominal SV onto the signal-plus-interference subspace to eliminate the arbitrary SV mismatch of SOI [14–17]. However, serious performance degradation will appear at low-input SNRs. In [15], a modified ESB method based on covariance matrix enhancement was proposed to improve the performance at low SNRs. In addition, the authors in [17] proposed a method to obtain the basis vector of signal subspace among the eigenvectors of the SCM. However, these methods perform poorly in the presence of large SV mismatch and high-input SNRs.

The uncertain-set-based technique utilizes a spherical or ellipsoidal uncertainty constraint setting on the nominal SV to estimate the SOI SV including the worst-case-based (WCB) method [5,18], doubly constrained method [10,19], probabilistically constrained method [20,21], and linear programming method [22]. However, these methods do not eliminate the SOI component from the SCM, and severe performance degradation will occur in the presence of high-input SNRs [7]. In addition, most of them need to solve the second-order cone programming (SOCP) problem, which leads to high complexity. Actually, the uncertain-set-based technique has been demonstrated to be equivalent to the DL method [6].

The above methods are mainly aimed at estimating SOI SV or SCM. Although these methods can improve the robustness of a beamformer, all still suffer from serious performance degradation at high-input SNRs. In order to overcome this drawback, a new type of RAB method based on INCM reconstruction has been developed in recent years. The authors of [23] firstly employ the SCB to estimate interference SV and reconstruct the INCM, but the power of interference was not accurately estimated. Gu, in [24], proposed an RAB method based on INCM reconstruction and SV estimation, where the INCM is reconstructed by integrating over the complement of the SOI angular region. However, the complexity is increased significantly. Subsequently, in [25,26], low-complexity shrinkage-based mismatch estimation (LOCSME) and the sparsity of the source distribution were used to significantly reduce the complexity. Unfortunately, these methods can achieve good performance only in certain conditions. To resist more types of mismatches, a new estimator for INCM based on interference SV and power estimation is presented in [6], and a QCQP problem with new inequality constraint was established to estimate the SOI SV. In [27], the authors constructed and solved a set of linear equations to obtain the estimation of interference power. Furthermore, the residual noise power was considered to improve the estimation accuracy of incident signal power in [28]. In [29], the iterative mismatch approximation method was employed to estimate the power and SV of all incident signals; then, these estimates were used to reconstruct the INCM. In [30], all nominal SVs were adjusted to an accurate version by a line search along the corresponding gradient vector. Together with the recorded power, the INCM was reconstructed. The Capon spectrum can be approximated as the power of noise when the SV mismatch is large enough. To overcome this drawback, the authors in [31] used the principle of maximum entropy power spectrum to reconstruct the interference and SOI covariance matrix by estimating all powers of incident signals. Different from the above INCM reconstruction-based methods, the authors in [32] reconstructed the INCM by projecting the interference subspace onto the received snapshots, which can effectively eliminate the SOI component and achieve good performance. In [33], the INCM reconstruction relies on using the average value of noise eigenvalues instead of the eigenvalue of the SOI to eliminate a noticeable part of the SOI. Ai et al. [34] presented an RAB algorithm for subspace projection and covariance reconstruction (SPCMR) that employs subspace projection and oblique projection to estimate the SOI SV and interference powers accurately. In [35], each SV was derived from the vector located at the intersection of two subspaces. Meanwhile, the estimate of each SV was given in a closed-form expression.

In this paper, a low-complexity RAB method based on INCM reconstruction and SOI SV estimation is proposed. Unlike previous methods, the INCM in the proposed method was reconstructed utilizing the orthogonality of subspace. First, based on the idea of the matrix filter in [36,37], the minimum mean square error criterion was employed to construct the blocking matrix. Then, we performed eigen-decomposition on the SCM and obtained the orthogonal projection matrix of the signal subspace. By projecting the blocking matrix onto the orthogonal projection matrix, a projection matrix was obtained. Subsequently, the INCM was reconstructed by replacing the eigenvector columns of the SCM, which can span to the signal subspace with the corresponding eigenvectors of the projection matrix. Finally, the SOI SV was estimated by employing the iterative mismatch approximation method presented in [29]. The theoretical analysis and simulation results demonstrated that the proposed method can efficiently deal with multiple types of mismatches.

The rest of this paper is organized as follows. The signal model and necessary background regarding the adaptive beamforming method is introduced in Section 2. In Section 3, the proposed RAB methods are described in detail, and the feasibility analysis of the blocking matrix is performed. The simulation results are provided in Section 4. Finally, conclusions are drawn in Section 5.

2. Signal Model and Background

Consider a uniform linear array (ULA) composed of M omnidirectional sensors that are illuminated by $L + 1$ far-field uncorrelated narrowband signals, which consist of one SOI and L interferences. The array complex sample vector at time k can be presented as:

$$x(k) = x_s(k) + x_i(k) + x_n(k), \tag{1}$$

where $x_s(k) = s_0(k)a_0$ and $x_i(k) = \sum_{l=1}^{L} s_l(k)a_l$, respectively, represent the $M \times 1$ vector of the SOI and interference signal component in the received data. $s_l(k)$ and $a_l(l = 0, \ldots, L)$ are the lth incident signal waveform and corresponding SV. $x_n(k)$ is additive complex Gaussian noise with a zero mean and a variance of σ_n^2, which is uncorrelated with all the other signals. In this paper, the sensors were spaced at half of the wavelength. The nominal SV from θ can be written as:

$$\begin{aligned}
a(\theta) &= \left[1, e^{-j\frac{2\pi d}{\lambda}\sin\theta}, \ldots, e^{-j(M-1)\frac{2\pi d}{\lambda}\sin\theta}\right]^T \\
&= \left[1, e^{-j\pi\sin\theta}, \ldots, e^{-j(M-1)\pi\sin\theta}\right]^T,
\end{aligned} \tag{2}$$

where $(\cdot)^T$ denotes the transpose, and λ and d, respectively, denote signal wavelength and distance between two adjacent sensors.

The output of the beamformer is written as:

$$y(k) = w^H x(k), \tag{3}$$

where $w = (w_1, \ldots, w_M)^T$ is the complex beamformer weight vector, and $(\cdot)^H$ is the Hermitian transpose. The optimal beamformer weight vector, w, can be calculated by maximizing the output SINR, which is defined as follow:

$$SINR \triangleq \frac{\sigma_0^2 |w^H a_0|^2}{w^H R_{i+n} w}, \tag{4}$$

where $\sigma_0^2 = E\left[|s_0(k)|^2\right]$ and a_0 denote the SOI power and SV, $E[\cdot]$ denotes the expectation operator of the stochastic variable. $R_{i+n} \in \mathbb{C}^{M \times M}$ denotes the precise INCM which can be written as:

$$
\begin{aligned}
\boldsymbol{R}_{i+n} &= E\left\{[\boldsymbol{x}_i(k)+\boldsymbol{x}_n(k)][\boldsymbol{x}_i(k)+\boldsymbol{x}_n(k)]^H\right\}\\
&= \sum_{l=1}^{L}\sigma_l^2 \boldsymbol{a}_l \boldsymbol{a}_l^H + E\left[\boldsymbol{x}_n(k)\boldsymbol{x}_n^H(k)\right]\\
&= \boldsymbol{R}_i + \sigma_n^2\boldsymbol{I},
\end{aligned}
\tag{5}
$$

where $\sigma_l^2 = E\left[|s_l(k)|^2\right]$ and $\boldsymbol{I}$ denote the lth interference power and identity matrix, respectively, and σ_n^2 is the noise power. The main purpose of the optimal beamformer is to maximize the output SINR while keeping the SOI undistorted at the same time, which is the so-called the minimum variance distortionless response (MVDR) problem [2]:

$$
\min_{w} w^H \boldsymbol{R}_{i+n} w \; s.t. w^H \boldsymbol{a}_0 = 1.
\tag{6}
$$

The optimal beamformer weight vector is given by:

$$
w_{\text{opt}} = \frac{\boldsymbol{R}_{i+n}^{-1}\boldsymbol{a}_0}{\boldsymbol{a}_0^H \boldsymbol{R}_{i+n}^{-1}\boldsymbol{a}_0}.
\tag{7}
$$

In practical applications, the precise INCM $\boldsymbol{R}_{i+n}$ is always unavailable, and it is usually replaced by SCM $\hat{\boldsymbol{R}}$:

$$
\hat{\boldsymbol{R}} = \frac{1}{K}\sum_{k=1}^{K} x(k)x^H(k),
\tag{8}
$$

where K denotes the number of snapshots. As K increases, $\hat{\boldsymbol{R}}$ converge to the actual one. It has been proved that replacing $\boldsymbol{R}_{i+n}$ by SCM $\hat{\boldsymbol{R}}$ does not change the optimal output SINR [2]. Substituting the actual SOI SV $\boldsymbol{a}_0$ by the nominal SV $\bar{\boldsymbol{a}}_0$ based on the known array structure, the optimal weight vector becomes the sample covariance inversion (SMI) beamformer:

$$
w_{SMI} = \frac{\hat{\boldsymbol{R}}^{-1}\bar{\boldsymbol{a}}_0}{\bar{\boldsymbol{a}}_0^H \hat{\boldsymbol{R}}^{-1}\bar{\boldsymbol{a}}_0}.
\tag{9}
$$

With the optimal weight vector, the Capon spatial power spectrum, is employed as a power estimator over all directions [28]:

$$
\begin{aligned}
\hat{P}(\theta) &= w_{SMI}^H \hat{\boldsymbol{R}} w_{SMI}\\
&= \frac{1}{\bar{\boldsymbol{a}}^H(\theta)\hat{\boldsymbol{R}}^{-1}\bar{\boldsymbol{a}}(\theta)},
\end{aligned}
\tag{10}
$$

where $\bar{\boldsymbol{a}}(\theta)$ is the nominal SV associated with $\theta \in (-90^\circ, 90^\circ)$.

3. Proposed Method

The main idea of the proposed method is to utilize the reconstructed INCM and the corrected SOI SV to derive the optimal weight vector. Depending on the minimum mean square error criterion, a blocking matrix is obtained. The orthogonality of the subspace is employed to derive the projection matrix, and the INCM is reconstructed by replacing the eigenvector columns of the SCM such that they can span to the signal subspace with the corresponding eigenvectors of the projection matrix. The optimal weight vector is obtained along with the SOI SV estimated by the iterative mismatch approximation method.

3.1. INCM Reconstruction

Different from Capon power spectrum integration and interference estimation based INCM reconstruction methods, we present a blocking matrix based on the matrix filter principle in [36,37], which is suitable for suppressing signals illuminating within a specific angular region. Consider a blocking matrix $G \in \mathbb{C}^{M \times M}$, the property of G can be described as:

$$G^H \bar{a}(\theta) = \begin{cases} \bar{a}(\theta), & \theta \in \Theta_p \\ 0, & \theta \in \Theta_s \end{cases}, \tag{11}$$

where Θ_s denotes the stopband angular region, and the direction of arrival (DOA) of the SOI lies in it. Θ_p denotes the passband angular region which contains the locations of interference. Sampling Θ_s and Θ_p uniformly with the N_s and N_p sampling points, the corresponding angular sequences can be presented as θ_p $(p = 1, \ldots, N_p)$ and θ_s $(s = 1, \ldots, N_s)$. The blocking matrix design problem based on the minimum mean square error criterion can be described as:

$$\min_{G} \| G^H P - \check{P} \|_F, \tag{12}$$

where $\|\cdot\|_F$ denotes Frobenius norm. $P = [\bar{a}(\Theta_p), \bar{a}(\Theta_s)] \in \mathbb{C}^{M \times (N_p + N_s)}$ denotes the nominal manifold matrix, and $\check{P} = [\bar{a}(\Theta_p), 0_{M \times N_s}] \in \mathbb{C}^{M \times (N_p + N_s)}$ denotes the desired manifold matrix. The solution to (12) can be found by taking the gradient of $F(G) = G^H P - \check{P}_F$ and making it equal to zero:

$$\begin{aligned} \frac{\partial F(G)}{\partial G} &= \frac{\partial}{\partial G} tr \left[\left(G^H P - \check{P} \right)^H \left(G^H P - \check{P} \right) \right]^{\frac{1}{2}} = 0 \\ G &= \left(P P^H \right)^{-1} P \check{P}^H. \end{aligned} \tag{13}$$

Performing eigen decomposition on $\hat{R}$ yields:

$$\begin{aligned} \hat{R} &= \sum_{i=1}^{M} \Gamma_i u_i u_i^H = U \Gamma U^H \\ &= U_s \Gamma_s U_s^H + U_n \Gamma_n U_n^H, \end{aligned} \tag{14}$$

where $\Gamma_1 \geq \Gamma_2 \geq \cdots \Gamma_{M-1} \geq \Gamma_M$ denotes the eigenvalues of $\hat{R}$ arranged in descending order. The minimum eigenvalue can be approximately considered as the estimation of the noise power $\tilde{\sigma}_n^2$ [6]. $u_i \in \mathbb{C}^{M \times 1}$ is an eigenvector associated with Γ_i. $U_s = (u_1, \ldots, u_{L+1}) \in \mathbb{C}^{M \times (L+1)}$ and $U_n = (u_{L+2}, \ldots, u_M) \in \mathbb{C}^{M \times (M-L-1)}$, respectively, denote the signal subspace eigenvectors and the noise subspace eigenvectors. $\Gamma_s = diag(\Gamma_1, \ldots, \Gamma_{L+1}) \in \mathbb{C}^{M \times (L+1)}$ and $\Gamma_n = diag(\Gamma_{L+2}, \ldots, \Gamma_M) \in \mathbb{C}^{M \times (M-L-1)}$ are diagonal matrices. According to the properties of the eigen subspace, we have:

$$span\{u_1, \ldots, u_{L+1}\} = span\{a_0, \ldots, a_L\}, \tag{15}$$

where $span\{u_1, \ldots, u_{L+1}\}$ denotes the spanned subspace generated by the vector group $\{u_1, \ldots, u_{L+1}\}$. Then, any accurate SV a_l can be expressed as a linear combination of columns of U_s [27], which means that the projection matrix $\Phi = U_s U_s^H G$ is orthogonal to a_0. $Q = \| \Phi^H \bar{a}(\theta) \|^2$ is used to measure the orthogonality between projection matrix Φ and $\bar{a}(\theta)$. Assume that the SOI and interference impinge on the half-wavelength spacing ULA with $M = 10$ from $\theta_0 = 3°$, $\theta_1 = -35°$, and $\theta_2 = 42°$, the stopband and passband region are set as $\Theta_s = (\theta_0 - 6°, \theta_0 + 6°)$ and $\Theta_p = (-90°, \theta_0 - 6°) \bigcup (\theta_0 + 6°, 90°)$. It can be observed from Figure 1 that Q will be much smaller when $\theta = \theta_0$ than $\theta = \theta_{1,2}$. This means Φ and $\bar{a}(\theta_0)$ are orthogonal or approximately orthogonal. In addition, the value of Q corresponding to $\theta_{1,2}$ is equal to $\|\bar{a}(\theta_{1,2})\|^2$. Hence, projection matrix Φ collects the spatial information of interference and removes the spatial information of the SOI.

Consider that the interference SV lies in the signal subspace spanned by the dominant eigenvectors of Φ. Then, employing eigen decomposition on Φ to obtain the signal subspace:

$$\Phi = B \Lambda B^H = B_s \Lambda_s B_s^H + B_n \Lambda_n B_n^H, \tag{16}$$

where $\boldsymbol{B} = [\boldsymbol{b}_1, \ldots, \boldsymbol{b}_M] = [\boldsymbol{B}_s, \boldsymbol{B}_n]$ and $\boldsymbol{\Lambda} = diag\{\lambda_1, \ldots, \lambda_M\}$, respectively, denote unitary and diagonal matrices. $\boldsymbol{b}_l, l = 1, \ldots, M$ denotes the eigenvector corresponding to λ_l, which are arranged in descending order. In addition, $\boldsymbol{B}_s$ contains $L + 1$ eigenvectors columns corresponding to the $L + 1$ largest eigenvalues, and $span\{\boldsymbol{b}_1, \ldots, \boldsymbol{b}_{L+1}\}$ can be considered as the signal subspace.

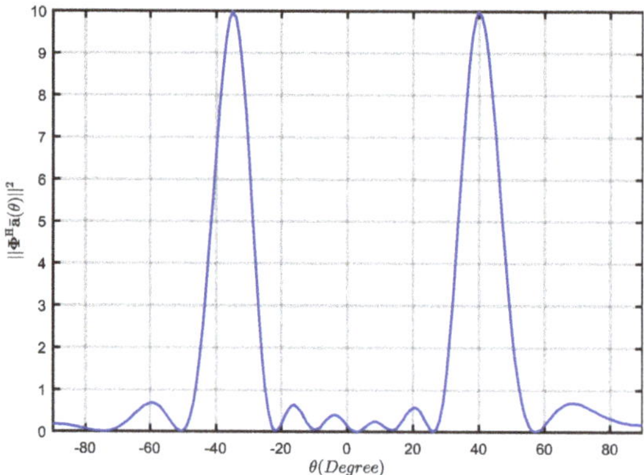

Figure 1. The value of $\|\boldsymbol{\Phi}^H \overline{\boldsymbol{a}}(\theta)\|^2$ versus θ.

Substituting $\boldsymbol{B}_s$ back into (14) and replacing $\boldsymbol{U}_s$, then we can obtain the reconstructed INCM:

$$\widetilde{\boldsymbol{R}}_{i+n} = \boldsymbol{B}_s \boldsymbol{\Gamma}_s \boldsymbol{B}_s^H + \boldsymbol{U}_n \boldsymbol{\Gamma}_n \boldsymbol{U}_n^H, \tag{17}$$

plotting the power spectrum of $\widetilde{\boldsymbol{R}}_{i+n}$ by replacing $\widetilde{\boldsymbol{R}}_{i+n}$ with $\hat{\boldsymbol{R}}$ in (10), which is written as:

$$\widetilde{P}(\theta) = \frac{1}{\overline{\boldsymbol{a}}^H(\theta) \widetilde{\boldsymbol{R}}_{n+i}^{-1} \overline{\boldsymbol{a}}(\theta)}. \tag{18}$$

The spatial power spectrum distribution based on (10) and (18) is drawn in Figure 2. The SOI was assumed to be impinging from $\theta_0 = 3°$ with a fixed $SNR = 30$ dB and that two interferences were impinging from $\theta_1 = -35°$ and $\theta_2 = 42°$ with a fixed interference-to-noise rate (INR) $INR = 20$ dB, respectively. This shows that the blocking matrix can effectively filter the SOI components and interference components are retained. Therefore, $\widetilde{\boldsymbol{R}}_{i+n}$ can be used as an INCM to derive the beamformer.

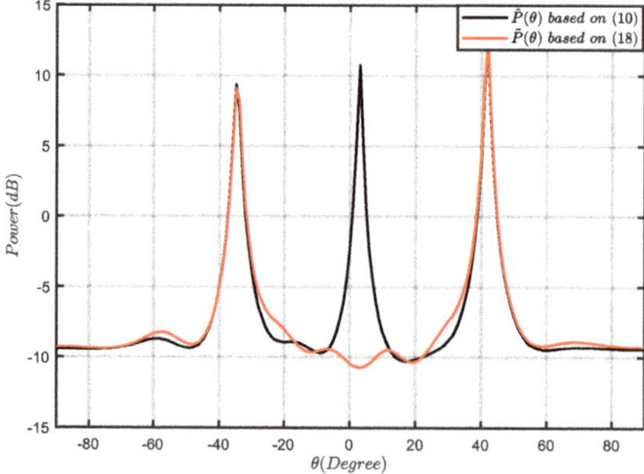

Figure 2. Comparison of the Capon power spectrum (10) and the power spectrum based on (18) with $SNR = 30$ dB, $INR = 20$ dB, and $K = 100$.

3.2. SOI SV Estimation and Beamformer Weight Vector Calculation

For estimating the SOI SV, we employed the iterative mismatch approximation method proposed in [29] to correct the presumed SOI SV. The iterative mismatch approximation method depends on searching for the SV mismatch in the margin of the amplitude and phase error. In the presence of SV mismatches, the actual SOI SV can be written as $a_0 = \bar{a}_0 + e = \alpha \circ \bar{a}_0 \circ e^{j\beta}$. Employing the principle of estimating the signal steering vectors via maximizing the beamformer output power [16], the estimated SOI SV $\tilde{a}_0$ can be obtained by solving the pseudo-optimization problem:

$$\begin{aligned} \max_{\alpha,\beta} \quad & \tilde{P}(\theta_q, \alpha, \beta) \\ s.t. \quad & |1 - \alpha_m| \leq \epsilon \\ & |\beta_m| \leq \Phi, \end{aligned} \tag{19}$$

where $\circ$ denotes the Hadamard product, and the power spectrum associated with θ_q, α, β is written as $\tilde{P}(\theta_q, \alpha, \beta) = 1/(\alpha \circ \bar{a}(\theta_q) \circ e^{j\beta})^H \hat{R}^{-1}(\alpha \circ \bar{a}(\theta_q) \circ e^{j\beta})$. ϵ and Φ denote the predefined boundary values of the amplitude and phase mismatch, respectively. $\alpha = (\alpha_1, \ldots, \alpha_M)^T$ and $\beta = (\beta_1 \ldots, \beta_M)^T$ denote the amplitude and phase mismatch vectors, respectively. The iterative mismatch approximation method is described in Algorithm 1. In addition, the initial nominal SOI SV is associated with the middle value of Θ_s.

Algorithm 1. Iterative mismatch approximation method.

Input: ϵ, Φ, $\bar{a}_q$, $\hat{R}^{-1}$
Output: $\tilde{a}_q$, pwr
1:    Initialize $a = \bar{a}_q$
2:    **for** $it1 = 1 \ldots M$
3:       $\tilde{\Theta}_\alpha = \left[\alpha_{ql}, \alpha_{qh}\right] = [1 - \epsilon,\ 1 + \epsilon],\ \tilde{\Theta}_\beta = \left[\beta_{ql}, \beta_{qh}\right] = \left[e^{j(-\Phi)}, e^{j(\Phi)}\right]$
4:       **for** $it2 = 1 \ldots depth$
5:          $E_{ap} = \tilde{\Theta}_\beta \otimes \tilde{\Theta}_\alpha,\ E_{ap} \in \mathbb{C}^{1 \times 4}$
6:          $a_i(it1) = \bar{a}_q(it1)E_{gp}(i), i = 1, \ldots, 4$
7:          Calculate $p(i)$ by substituting a_i into Equation (10)
8:          $(pwr, idx) = max(p)$
9:          Zoom out of the amplitude/phase error area built by $\tilde{\Theta}_\alpha$ and $\tilde{\Theta}_\beta$ to $E_{ap}(idx)$
10:        Update $\tilde{\Theta}_\alpha$ and $\tilde{\Theta}_\beta$
11:       **end**
12:       $\tilde{a}_q(it1) = \bar{a}_q(it1)E_{ap}(idx)$
13: **end**

Then, the estimated SOI SV $\tilde{a}_0$ can be corrected as $\tilde{a}_q$. Substituting (17) together with $\tilde{a}_0$ back into the Capon beamformer (7), the robust adaptive beamforming based on INCM reconstruction via the projecting matrix and SV estimation can be written as:

$$\tilde{w} = \frac{\tilde{R}_{i+n}^{-1}\tilde{a}_0}{\tilde{a}_0^H \tilde{R}_{i+n}^{-1}\tilde{a}_0}. \tag{20}$$

In a general case, the estimated SOI SV may be imprecise. Hence, we took $\tilde{a}_0$ as an input parameter and performed multiple iterations to improve accuracy. The iteration was terminated when the following conditions were satisfied:

$$\left|\frac{\tilde{\sigma}_q^2\big|_{current} - \tilde{\sigma}_q^2\big|_{previous}}{\tilde{\sigma}_q^2\big|_{previous}}\right| < \varphi, \tag{21}$$

where φ and $\widetilde{\sigma}_q^2$ denote a predefined saturation coefficient and power corresponding to $\widetilde{a}_q$, respectively. The method we propose is summarized in Algorithm 2.

Algorithm 2. Proposed RAB method.

1: Calculate the SCM $\hat{R}$ using (8) and eigen decompose $\hat{R}$ to obtain U_S and $\widetilde{\sigma}_n^2$;
2: Obtain the blocking matrix G using (13) and the projection matrix $\boldsymbol{\Phi} = U_s U_s^H G$;
3: Eigen decompose $\boldsymbol{\Phi}$ to obtain B_s and reconstruct INCM via (17);
4: Using the iterative mismatch approximation method in Algorithm 1 to estimate the SOI SV;
5: Substitute $\widetilde{R}_{i+n}$ and $\widetilde{a}_0$ back into (20) to obtain the weight vector.

The number of floating-point operations (flops) was employed to measure the computational complexity. The computational complexity of the proposed method is mainly determined by calculating the SCM $\hat{R}$ and matrix eigen decomposition, calculating G and estimating SOI SV $\widetilde{a}_0$. Calculating the SCM $\hat{R}$ costs approximately $O(M^2 K)$ flops. Calculating G costs approximately $O(N_S M^2 + N_P M^2 + 2M^3)$ flops. The computational complexity of matrix inversion and eigen decomposition is approximately $O(M^3)$ flops. Assuming that $depth = D$, the computational complexity of estimating the SOI SV would be approximately $O(M^3 + 3DM)$ flops. In practice, calculating G is independent of $\hat{R}$ and can be seen as a pre-processing operation of the proposed method. With $K \gg M \cong D$, the overall complexity of proposed method is approximately $O(M^2 K)$ flops. In [6], the major computations were conducted to solve a QCQP problem and estimate interference SVs. Suppose that S denotes the number of search points in $\overline{\Theta}$, the computational complexity is $O(max(M^2 S, M^{3.5}))$. In [30], the sample points, N, for the line search significantly affected the adjustment of the SVs. Hence, when $N > M$, the computational complexity is $O(LNM^2)$. The computational complexity of the MEPS-IPNC in [31] was $O(M^2 L)$, where L is a small multiple of M. In [32], the computational complexity was $O(max(M^2 J, M^2 S))$, where S and J are the number of the sampling points in Θ_s and $\overline{\Theta_s \cup \Theta_i}$. In [33], the computational complexity was declared to be $O(M^3)$. In [34], the computational complexity was $O(max(M^2 J, M^2 S))$, where S and J are the number of the sampling points in Θ and $\overline{\Theta}$. In [35], the computational complexity was declared to be $O(M^2 I(L+1))$, where I is the number of sample points for Θ and Θ_{i-l}. Clearly, since the power spectrum calculation, power spectrum integration, and QCQP problem solving are avoided in our proposed method, it has the lowest computational complexity. Furthermore, prior information regarding the SOI angular region and array geometry are needed.

4. Simulation

In this section, a half-wavelength spacing ULA with $M = 10$ was considered. We assumed that there existed an SOI impinging from the direction of $\theta_0 = 3°$ and two interferences impinging from $\theta_1 = -35°$ and $\theta_2 = 42°$. The additive noise was presumed to be a complex circularly symmetric Gaussian zero-mean unit-variance spatially and temporally white process. All these sources were narrowband and assumed to be independent to the noise. To obtain each output SINR point, 200 Monte Carlo trials were used in each simulation. The proposed method was compared with the RAB method based on INCM reconstruction and steering vector estimation (INCM-SVE) in [6], subspace-decomposition and SV adjustment (SDA) in [30], MEPS-IPNC in [31], INCM reconstruction via orthogonality of subspace (INCM-OS) in [32], desired signal eigenvalue replacement (DSEB) in [33], SPCMR in [34] and INCM reconstruction via the intersection of subspaces (INCM-IS) in [35]. For all methods involved in the comparisons, the angular region was presumed to be $\Theta = \Theta_s = (\theta_0 - 6°, \theta_0 + 6°)$ and $\overline{\Theta} = \Theta_p = (-90°, \theta_0 - 6°) \cup (\theta_0 + 6°, 90°)$. and the interference angular region to be $\Theta_i = (\theta_i - 6°, \theta_i + 6°)$. The number of non-dominant eigenvectors of the matrix C was set as $L = 7$, and the RCB boundary was $\epsilon = \sqrt{0.1}$ in [6]. The $N = 7$ dominant eigenvectors of matrix B were employed for B_1 in [32]. $\xi = 0.95$ as in [33]. The constant satisfying $\mu = 0.9$ and $\tau = \sqrt{0.1}$ were as in [34]. Sampling points

$L = 5M$ and $S = 20$ were as in [31]. The scale factor $\widetilde{\mu} = 0.1$ and the sampling points $N = 2\widetilde{\mu}/0.01$ were as in [30]. For the proposed method, the amplitude and phase mismatch boundary were set as $\epsilon = 0.3$ and $\Phi = 6°$, respectively, the depth of the iteration was set as $depth = 10$, and the saturation value was $\varphi = 0.05$. Furthermore, the MATLAB CVX toolbox was used to solve the QCQP optimization problem in [6]. In our simulations, the optimal output SINR can be calculated by:

$$SINR_{opt} = \sigma_0^2 \boldsymbol{a}_0^H \boldsymbol{R}_{i+n}^{-1} \boldsymbol{a}_0. \tag{22}$$

4.1. Example 1: Mismatch Due to the Amplitude and Phase Error of the SV

In the first example, the influence of the SVs with arbitrary amplitude and phase errors on the beamformer output SINR was considered. The relationship between the mth element of the nominal SV and the actual SV was modeled as $a_m = \alpha_m \bar{a}_m e^{j\beta_m}$, where the arbitrary amplitude error, α_m, and phase error, β_m, on each array sensor, respectively, followed the Gaussian distribution $N\left(1, 0.05^2\right)$ and $N\left(0, (5°)^2\right)$ [6]. Figure 3a depicts the output SINR of the tested methods versus the input SNR for the fixed number of snapshots $K = 100$. It was observed that the proposed method had a similar performance among the tested methods except in [32,33] at high SNRs. In addition, the performance of the proposed method was only lower than in [6] when the SNR was low. However, the computational complexity of our method was obviously lower than that in [6]. In Figure 3b, the output SINRs are shown versus the number of snapshots for the fixed $SNR = 30$ dB and $INR = 20$ dB. The proposed method had a similar performance to the tested methods in [6,30,34,35], and the number of snapshots did not affect the performance of our proposed method.

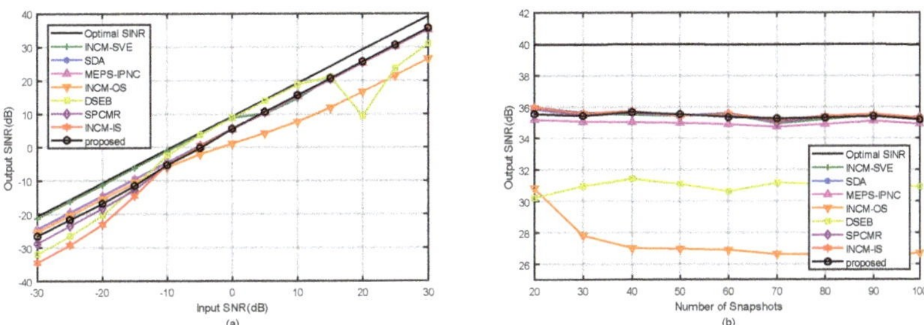

Figure 3. Output SINRs in the case of amplitude and phase errors versus (**a**) input SNR with $K = 100$; (**b**) the number of snapshots with $SNR = 30$ dB.

4.2. Example 2: Mismatch Due to the Random Look Direction Error

In the second example, the influence of the random look direction errors on the beamformer output SINR was considered. Assuming that the look direction mismatch of both the SOI and interferences were uniformly distributed in $(-5°, 5°)$. That is means that the DOA of the SOI was uniformly distributed in $(-2°, 8°)$, and the DOAs of the two interference were uniformly distributed in $(-40°, 30°)$ and $(37°, 47°)$. Note that the random DOAs of the SOI and interferences changed in each trial while remaining constant over snapshots. Figure 4a shows the output SINRs of the tested methods versus the input SNRs with the fixed snapshots $K = 100$. It was observed that our proposed method was only inferior to that in [6] in the performance at low SNR and inferior to that in [6,35] at high SNRs. Figure 4b depicts the output SINRs of the tested methods against the snapshot number at $SNR = 30$ dB and $INR = 20$ dB. It was observed that the performance of our proposed method was similar to that in [30] when $K > 40$. In addition, the methods in [33,34] were significantly affected by mismatches due to the look direction error.

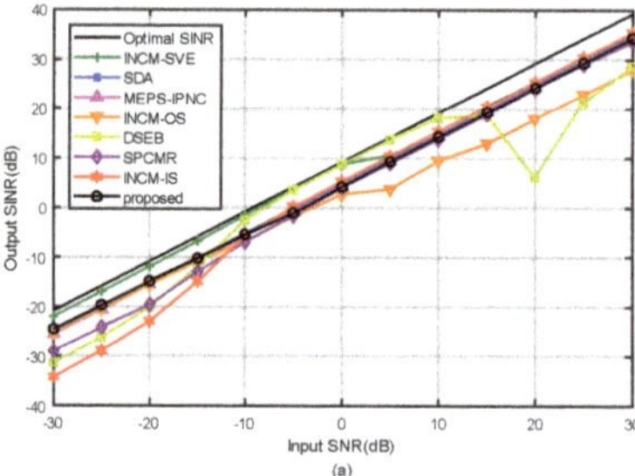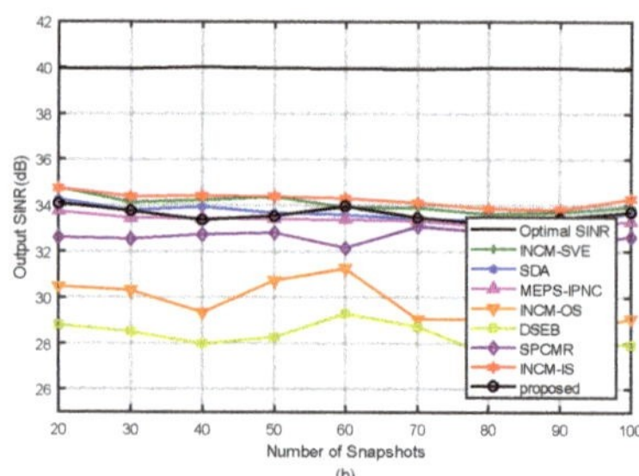

Figure 4. Output SINRs in the case of the look direction error versus (**a**) input SNRs with $K = 100$; (**b**) the number of snapshots with $SNR = 30$ dB.

4.3. Example 3: Mismatch Due to the Incoherent Local Scattering Error

In the third example, the influence of the incoherent local scattering error on the beamformer output SINRs was considered. The SOI was assumed to have a time-varying signature, which was modeled as:

$$\hat{x}_s(k) = s_0(k)\boldsymbol{a}_0 + \sum_{p=1}^{4} s_p(k)\overline{\boldsymbol{a}}(\theta_p),\tag{23}$$

where $\boldsymbol{a}_0$ denotes the SOI SV. $\overline{\boldsymbol{a}}(\theta_p)\,(p = 1, 2, 3, 4)$ denotes the incoherent scattering signal SV, and the DOAs, θ_p, are independently distributed in a Gaussian distribution drawn from a random generator $N(\theta_0, 4°)$ in each trial. $s_p(k)$ are independently and identically distributed zero-mean complex Gaussian random variables independently drawn from a random generator, $N(0, 1)$. In addition, θ_p changes from trial to trial, while it remains fixed over the samples. At the same time, $s_p(k)$ changes both from trial to trial and from sample to sample. In this case, the SOI covariance matrix is no longer a rank-one matrix and the output SINR should be expressed as [5]:

$$SINR_{opt} = \frac{\boldsymbol{w}^H \boldsymbol{R}_s \boldsymbol{w}}{\boldsymbol{w}^H \boldsymbol{R}_{i+n} \boldsymbol{w}}.\tag{24}$$

The optimal weight vector can be obtained by maximizing the SINR [5]:

$$\boldsymbol{w}_{opt} = P\left\{\boldsymbol{R}_{i+n}^{-1} \boldsymbol{R}_s\right\},\tag{25}$$

where $P\{\cdot\}$ denotes the principal eigenvector of a matrix. Figure 5a shows the output SINRs of the tested methods versus the input SNRs with the fixed snapshots $K = 100$. It was observed that the performance of our proposed method was similar to that in [6,30] at high SNRs and only lower than in [33] at low SNRs. However, the method in [33] had severe performance degradation at high SNRs. Figure 5b depicts the output SINRs of the tested methods against the snapshot number at $SNR = 30$ dB and $INR = 20$ dB. It was observed that the proposed method had a similar performance with the tested methods in [6,30,35], and the number of snapshots did not affect the performance of our proposed method.

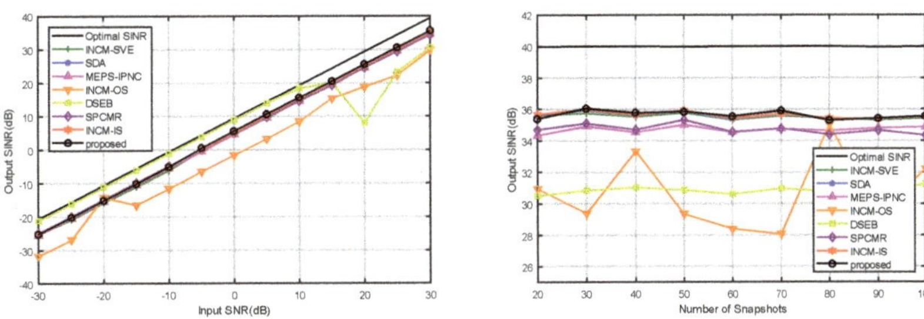

Figure 5. Output SINRs in the case of incoherent local scattering error versus (**a**) input SNRs with $K = 100$; (**b**) the number of snapshots with $SNR = 30$ dB.

4.4. Example 4: Mismatch Due to the Coherent Local Scattering Error

In the fourth example, the influence of the coherent local scattering mismatch on the beamformer output SINRs was considered. The coherent local scattering mismatch usually occurs in multipath propagation scenarios. Assume that the SOI is distorted by local scattering and consists of four coherent paths; the actual SV is formed as:

$$\hat{a}_0 = a_0 + \sum_{p=1}^{4} e^{j\Phi_p}\overline{a}(\theta_p), \tag{26}$$

where a_0 denotes the SOI SV. $\overline{a}(\theta_p)\,(p = 1, 2, 3, 4)$ denotes the coherent signal path from θ_p. θ_p are independently distributed in a Gaussian distribution drawn from a random generator, $N(\theta_0, 4°)$, in each trial. Φ_p denotes the path phase and uniformly distributed in $(0, 2\pi)$ from trial to trial. θ_p and Φ_p change from trial to trial, while it remains fixed over the samples. Figure 6a shows the output SINRs of the tested methods versus the input SNRs with the fixed snapshots $K = 100$. It was observed that the performance of the optimal beamformer had an approximately 6 dB increment in output SINRs due to the extra paths. The performance of our proposed method was similar to that in [6,30,35] at high SNRs. The method in [6] achieved the best performance at the cost of the highest complexity compared with the others. Figure 6b depicts the output SINRs of the tested methods against the snapshot number at $SNR = 30$ dB and $INR = 20$ dB. It was observed that the proposed method had a small impact on the number of snapshots.

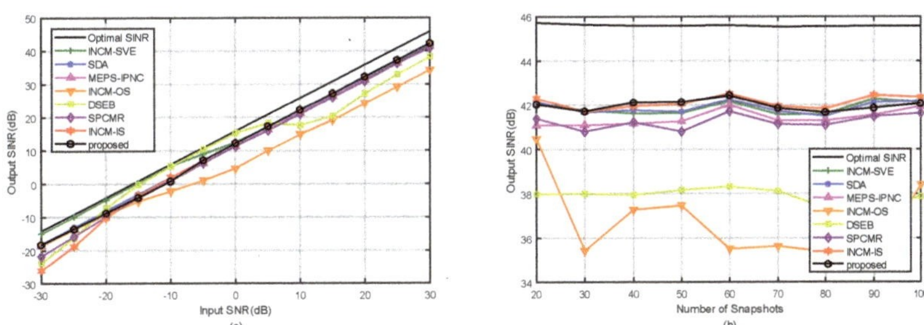

Figure 6. Output SINRs in the case of coherent local scattering error versus (**a**) input SNR with $K = 100$; (**b**) the number of snapshots with $SNR = 30$ dB.

5. Conclusions

In this paper, we proposed a low-complexity RAB method based on INCM reconstruction via subspace projection. In this method, the component of the SOI in the SCM was eliminated by replacing the eigenvector columns in the SCM such that they could span to the signal subspace with the corresponding eigenvectors in the projection matrix.

Meanwhile, the SOI SV was estimated by employing the iterative mismatch approximation method. Since the calculation of the blocking matrix can be seen as pre-processing, the complexity of INCM reconstruction only depends on a countable number of matrices' eigen decomposition and multiplication. Both the analysis and simulation illustrate that the proposed method is robust to various types of mismatches while maintaining low complexity.

Author Contributions: All authors have significantly contributed to the research presented in this manuscript. Y.D. presented the main idea and wrote the manuscript. X.Y. edited the manuscript. L.M. and W.C. reviewed and revised the manuscript. All authors have read and agreed to the published version of the manuscript.

Funding: This research was funded by the Laboratory of Science and Technology on Information Transmission and Dissemination in Communication Networks under Grant SXX19641X071.

Institutional Review Board Statement: Not applicable.

Informed Consent Statement: Not applicable.

Data Availability Statement: Not applicable.

Conflicts of Interest: The authors declare no conflict of interest.

References

1. Reed, I.S.; Mallett, J.D.; Brennan, L.E. Rapid Convergence Rate in Adaptive Arrays. *IEEE Trans. Aerosp. Electron. Syst.* **1974**, *AES-10*, 853–863. [CrossRef]
2. Li, J.; Stoica, P. *Robust Adaptive Beamforming/Edited by Jian Li and Petre Stoica*; John Wiley: Hoboken, NJ, USA, 2006; p. xii. 422p.
3. Van Trees, H.L.; Bell, K.L.; Tian, Z. *Detection Estimation and Modulation Theory*, 2nd ed.; John Wiley & Sons, Inc.: Hoboken, NJ, USA, 2013.
4. Du, L.; Li, J.; Stoica, P. Fully Automatic Computation of Diagonal Loading Levels for Robust Adaptive Beamforming. *IEEE Trans. Aerosp. Electron. Syst.* **2010**, *46*, 449–458. [CrossRef]
5. Vorobyov, S.A.; Gershman, A.B.; Luo, Z.Q. Robust adaptive beamforming using worst-case performance optimization: A solution to the signal mismatch problem. *IEEE Trans. Signa Process.* **2003**, *51*, 313–324. [CrossRef]
6. Zheng, Z.; Zheng, Y.; Wang, W.Q.; Zhang, H.B. Covariance Matrix Reconstruction With Interference Steering Vector and Power Estimation for Robust Adaptive Beamforming. *IEEE Trans. Veh. Technol.* **2018**, *67*, 8495–8503. [CrossRef]
7. Zheng, Z.; Yang, T.; Wang, W.Q.; So, H.C. Robust Adaptive Beamforming via Simplified Interference Power Estimation. *IEEE Trans. Aerosp. Electron. Syst.* **2019**, *55*, 3139–3152. [CrossRef]
8. Cox, H.; Zeskind, R.; Owen, M. Robust adaptive beamforming. *IEEE Trans. Acoust. Speech Signal Process.* **1987**, *35*, 1365–1376. [CrossRef]
9. Carlson, B.D. Covariance matrix estimation errors and diagonal loading in adaptive arrays. *IEEE Trans. Aerosp. Electron. Syst.* **1988**, *24*, 397–401. [CrossRef]
10. Beck, A.; Eldar, Y.C. Doubly constrained robust capon beamformer with ellipsoidal uncertainty sets. *IEEE Trans. Signal Process.* **2007**, *55*, 753–758. [CrossRef]
11. Stoica, P.; Li, J.; Zhu, X.M.; Guerci, J.R. On using a priori knowledge in space-time adaptive processing. *IEEE Trans. Signal Process.* **2008**, *56*, 2598–2602. [CrossRef]
12. Yang, J.; Ma, X.C.; Hou, C.H.; Liu, Y.C. Automatic Generalized Loading for Robust Adaptive Beamforming. *IEEE Signal Process. Lett.* **2009**, *16*, 219–222. [CrossRef]
13. Zhang, M.; Chen, X.M.; Zhang, A.X. A simple tridiagonal loading method for robust adaptive beamforming. *Signal Process.* **2019**, *157*, 103–107. [CrossRef]
14. Feldman, D.D.; Griffiths, L.J. A Projection Approach for Robust Adaptive Beamforming. *IEEE Trans. Signal Process.* **1994**, *42*, 867–876. [CrossRef]
15. Huang, F.; Sheng, W.X.; Ma, X.F. Modified projection approach for robust adaptive array beamforming. *Signal Process.* **2012**, *92*, 1758–1763. [CrossRef]
16. Jia, W.M.; Jin, W.; Zhou, S.H.; Yao, M.L. Robust adaptive beamforming based on a new steering vector estimation algorithm. *Signal Process.* **2013**, *93*, 2539–2542. [CrossRef]
17. Huang, L.; Zhang, B.; Ye, Z.F. Robust Adaptive Beamforming Using a New Projection Approach. In Proceedings of the 2015 IEEE International Conference on Digital Signal Processing (DSP), Singapore, 21–24 July 2015; pp. 1181–1185.
18. Yu, Z.L.; Gu, Z.H.; Zhou, J.J.; Li, Y.Q.; Ser, W.; Er, M.H. A Robust Adaptive Beamformer Based on Worst-Case Semi-Definite Programming. *IEEE Trans. Signal Process.* **2010**, *58*, 5914–5919. [CrossRef]
19. Li, J.; Stoica, P.; Wang, Z.S. Doubly constrained robust Capon beamformer. *IEEE Trans. Signal Process.* **2004**, *52*, 2407–2423. [CrossRef]

20. Vorobyov, S.A.; Rong, Y.; Gershman, A.B. Robust adaptive beamforming using probability-constrained optimization. In Proceedings of the IEEE/SP 13th Workshop on Statistical Signal Processing, Bordeaux, France, 17–20 July 2005; pp. 869–874.
21. Vorobyov, S.A.; Chen, H.H.; Gershman, A.B. On the Relationship Between Robust Minimum Variance Beamformers With Probabilistic and Worst-Case Distortionless Response Constraints. *IEEE Trans. Signal Process.* **2008**, *56*, 5719–5724. [CrossRef]
22. Jiang, X.; Zeng, W.J.; Yasotharan, A.; So, H.C.; Kirubarajan, T. Robust Beamforming by Linear Programming. *IEEE Trans. Signal Process.* **2014**, *62*, 1834–1849. [CrossRef]
23. Mallipeddi, R.; Lie, J.P.; Suganthan, P.N.; Razul, S.G.; See, C.M.S. Robust adaptive beamforming based on covariance matrix reconstruction for look direction mismatch. *PIER Lett.* **2011**, *25*, 34–46. [CrossRef]
24. Gu, Y.J.; Leshem, A. Robust Adaptive Beamforming Based on Interference Covariance Matrix Reconstruction and Steering Vector Estimation. *IEEE Trans. Signal Process.* **2012**, *60*, 3881–3885. [CrossRef]
25. Gu, Y.J.; Goodman, N.A.; Hong, S.H.; Li, Y. Robust adaptive beamforming based on interference covariance matrix sparse reconstruction. *Signal Process.* **2014**, *96*, 375–381. [CrossRef]
26. Ruan, H.; de Lamare, R.C. Robust Adaptive Beamforming Using a Low-Complexity Shrinkage-Based Mismatch Estimation Algorithm. *IEEE Signal Process. Lett.* **2014**, *21*, 60–64. [CrossRef]
27. Zheng, Z.; Wang, W.Q.; So, H.C.; Liao, Y. Robust adaptive beamforming using a novel signal power estimation algorithm. *Digit. Signal Process.* **2019**, *95*, 102574. [CrossRef]
28. Zhu, X.Y.; Ye, Z.F.; Xu, X.; Zheng, R. Covariance Matrix Reconstruction via Residual Noise Elimination and Interference Powers Estimation for Robust Adaptive Beamforming. *IEEE Access* **2019**, *7*, 53262–53272. [CrossRef]
29. Duan, Y.L.; Zhang, S.L.; Cao, W.P. Covariance matrix reconstruction with iterative mismatch approximation for robust adaptive beamforming. *J. Electromagnet Waves Appl.* **2021**, *35*, 2468–2479. [CrossRef]
30. Sun, S.C.; Ye, Z.F. Robust adaptive beamforming based on a method for steering vector estimation and interference covariance matrix reconstruction. *Signal Process.* **2021**, *182*, 107939. [CrossRef]
31. Mohammadzadeh, S.; Nascimento, V.H.; Lamare, R.C.d.; Kukrer, O. Maximum Entropy-Based Interference-Plus-Noise Covariance Matrix Reconstruction for Robust Adaptive Beamforming. *IEEE Signal Process. Lett.* **2020**, *27*, 845–849. [CrossRef]
32. Zhu, X.Y.; Xu, X.; Ye, Z.F. Robust adaptive beamforming via subspace for interference covariance matrix reconstruction. *Signal Process.* **2020**, *167*, 107289. [CrossRef]
33. Shen, F.; Chen, F.F.; Song, J.Y. Robust Adaptive Beamforming Based on Steering Vector Estimation and Covariance Matrix Reconstruction. *IEEE Commun. Lett.* **2015**, *19*, 1636–1639. [CrossRef]
34. Ai, X.Y.; Gan, L. Robust Adaptive Beamforming With Subspace Projection and Covariance Matrix Reconstruction. *IEEE Access* **2019**, *7*, 102149–102159. [CrossRef]
35. Wang, R.; Wang, Y.; Han, C.; Gong, Y.; Wang, L. Robust Adaptive Beamforming Based on Interference Covariance Matrix Reconstruction and Steering Vector Estimation. In Proceedings of the 2021 IEEE International Conference on Signal Processing, Communications and Computing (ICSPCC), Xi'an, China, 17–19 August 2021; pp. 1–5.
36. Vaccaro, R.J.; Chhetri, A.; Harrison, B.F. Matrix filter design for passive sonar interference suppression. *J. Acoust. Soc. Am.* **2004**, *115*, 3010–3020. [CrossRef]
37. Yan, S.F.; Ma, Y.L. Optimal design and verification of temporal and spatial filters using second-order cone programming approach. *Sci. China Ser. F* **2006**, *49*, 235–253. [CrossRef]

Article

Efficient Design for Integrated Photonic Waveguides with Agile Dispersion

Zhaonian Wang, Jiangbing Du *, Weihong Shen, Jiacheng Liu and Zuyuan He

State Key Laboratory of Advanced Optical Communication Systems and Networks, Shanghai Jiao Tong University, Shanghai 200240, China; ee_wzn@sjtu.edu.cn (Z.W.); shenweihong@sjtu.edu.cn (W.S.); jiacheng.liu@sjtu.edu.cn (J.L.); zuyuanhe@sjtu.edu.cn (Z.H.)
* Correspondence: dujiangbing@sjtu.edu.cn

Abstract: Chromatic dispersion engineering of photonic waveguide is of great importance for Photonic Integrated Circuit in broad applications, including on-chip CD compensation, supercontinuum generation, Kerr-comb generation, micro resonator and mode-locked laser. Linear propagation behavior and nonlinear effects of the light wave can be manipulated by engineering CD, in order to manipulate the temporal shape and frequency spectrum. Therefore, agile shapes of dispersion profiles, including typically wideband flat dispersion, are highly desired among various applications. In this study, we demonstrate a novel method for agile dispersion engineering of integrated photonic waveguide. Based on a horizontal double-slot structure, we obtained agile dispersion shapes, including broadband low dispersion, constant dispersion and slope-maintained linear dispersion. The proposed inverse design method is objectively-motivated and automation-supported. Dispersion in the range of 0–1.5 ps/(nm·km) for 861-nm bandwidth has been achieved, which shows superior performance for broadband low dispersion. Numerical simulation of the Kerr frequency comb was carried out utilizing the obtained dispersion shapes and a comb spectrum for 1068-nm bandwidth with a 20-dB power variation was generated. Significant potential for integrated photonic design automation can be expected.

Keywords: dispersion engineering; slot waveguide; inverse design; deep neural network; optical frequency comb

Citation: Wang, Z.; Du, J.; Shen, W.; Liu, J.; He, Z. Efficient Design for Integrated Photonic Waveguides with Agile Dispersion. *Sensors* **2021**, *21*, 6651. https://doi.org/10.3390/s21196651

Academic Editor: Luis Velasco

Received: 8 September 2021
Accepted: 4 October 2021
Published: 7 October 2021

Publisher's Note: MDPI stays neutral with regard to jurisdictional claims in published maps and institutional affiliations.

1. Introduction

Photonic Integrated Circuit (PIC) is essential for integrated optical systems, and thus plays a key role in broad applications. In a PIC, the control of waveguide's chromatic dispersion is of central importance in most of applications, which significantly influences the propagation of light field and the generation of various nonlinearity effects [1]. Waveguides with agile shapes of dispersion profiles could potentially constitute various functional devices. The targets of dispersion engineering may include fiber-induced dispersion compensation, tunable dispersion for mode locking of lasers, as well as many other relevant devices [2–5]. For instance, the generation of a light soliton comb requires a flat and low anomalous dispersion [6–9], while with normal dispersion, dark soliton, can be generated for obtaining improved optical combs [10–13]. Meanwhile, constant or extremely large dispersion may also be useful in applications of optical phase array, on-chip FWM/OPA process, and optical delay line [14–18]. In this way, agile dispersion with various shapes are highly desired for various on-chip applications.

The dispersion applied in a photonic integrated circuit is mainly designed by changing the materials and structures of a waveguide. One of the common ways to shape the dispersion is to change the dimensions of its structures [19,20], and find the target structure. This is a process of forward design, which may be quite difficult to execute due to the large amount of multi-parameter sweeping of simulation and optimization for complex structures. More complex structures of waveguides usually support the potential for better performance of dispersion design. The increased parameter inevitably leads to higher calculation difficulties. Therefore, a key point of dispersion engineering for a waveguide is to find an appropriate structure accurately and efficiently that possesses the target dispersion profile. By considering this situation, machine learning might be a valid method to achieve this objective. Through proper training of the Neural Network, mapping between input and output coefficients can be solved. Then, multiple parameters can be optimized simultaneously. This kind of inverse design method has been applied to quite a lot of photonic structure and fiber designs [21–24], and was found to be competent for supporting photonic design automation (PDA).

This paper proposes a novel inverse design approach to solve agile dispersion engineering by using Neural Network. A horizontal double slot waveguide structure is adopted with As2S3 and Si3N4 utilized. After generation of a certain amount of data set, the inverse design method cost little time for completing optimization. This process provides better accuracy, efficiency, and lower-complexity for fast and agile dispersion design of waveguides with reusability for design automation. Moreover, it can be extended to other kinds of optical devices. Therefore, great potential for design automation of PIC can be expected.

2. Principle of Dispersion Inverse Design

2.1. Waveguide Structure for Inverse Design

For the inverse design in this work, a horizontal double slot waveguide structure was adopted. Figure 1a shows six structural parameters of H1, H2, H3, Hs1, Hs2, and W, which have low refractive index materials lying between high index materials and the area of low index materials is called slot. In this work, the materials, As_2S_3 and Si_3N_4, are employed as different high index materials, respectively, and SiO_2 is a low index material. The cladding is air. The material, As_2S_3, also named chalcogenide glass, has broadband infrared transparency and low linear and nonlinear propagation loss at C band and 2 μm band, and has been gradually applied for photonic integrated chips [25]. The material, Si_3N_4, is renowned for its extremely low propagation loss and relatively mature chip processing. Due to low TPA at C band, Si_3N_4 has been utilized among various nonlinear mechanisms. Two kinds of materials have broad applications and play important roles in photonic integration. In this way, universality can be verified by applying these materials to inverse design processes.

Moreover, slot structures have large degrees of freedom with multiple parameters for better dispersion engineering. They are able to confine intense light field in the slot area, which favors the generation of various nonlinear effects [26]. Furthermore, horizontal slots have greater fabrication tolerance due to smaller sidewall angle and less loss, compared with vertical slot [27]. In terms of double and single slot structures, the former could confine significant light field in a wider wavelength range, as shown in Figure 1b,c, which means greater potential to achieve agile dispersion in a broader waveband [28]. The wavelengths corresponding to maximum ratio of double- and single-slot are marked in the Figure by dot lines in Figure 1c.

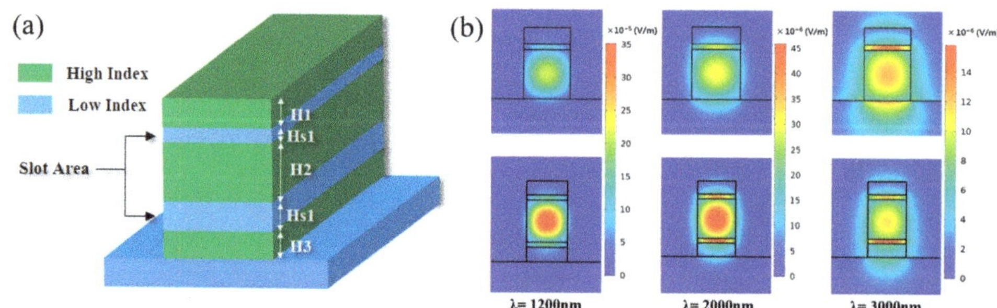

Figure 1. *Cont.*

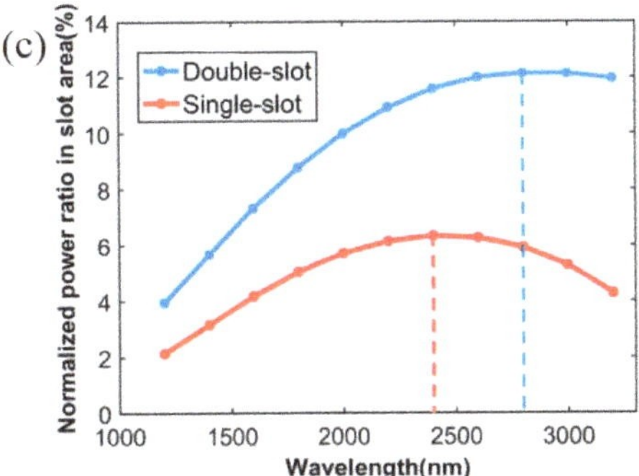

Figure 1. (**a**) Structures of horizontal double slot waveguide; (**b**) comparison of light field between single- and double-slot; (**c**): Comparison of the ratio for normalized power in slot area.

Considering all the reasons mentioned above, we chose this horizontal double slot waveguide with materials of As_2S_3 and Si_3N_4 as target objective of inverse design.

2.2. Data Generation and Inverse Design Process

The entire design process was aided by Neural Network for dispersion engineering, as shown in Figure 2. The main data required involved the dispersion of a large amount of waveguide structures, which was varied with wavelength, forming a dispersion profile. The dispersion profiles with respect to wavelength λ are calculated by the second derivative of the effective index n_{eff}, which can be obtained by Finite Element Method (FEM) simulation using commercial software (COMSOL for example). The relationship between n_{eff} and D can be described as follows in Equation (1):

$$D = -(\lambda/c) \cdot \left(\partial^2 n_{eff} / \partial \lambda^2 \right). \tag{1}$$

Among the simulation experiments, material dispersion has been taken into account by scientists [29], as illustrated by Sellmeier Equation with different values for As_2S_3 and Si_3N_4.

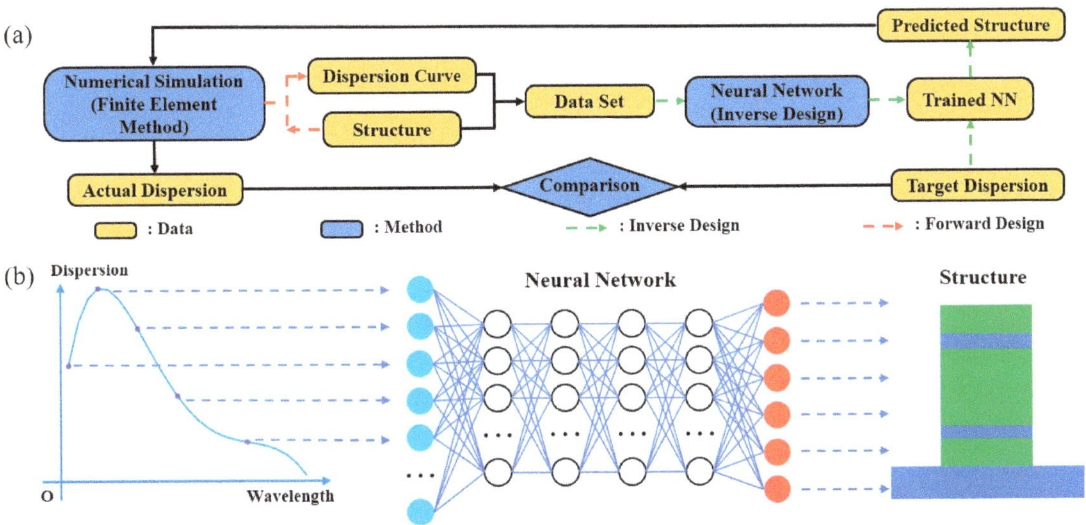

Figure 2. (**a**) Flow chart of the whole proposed inverse design method; (**b**) schematic principle of inverse design via Neural Network.

The above equation clarifies the method for obtaining dispersion profiles, and one important part of the Neural Network-based inverse design is to obtain a proper data set, which contains different structural parameters of waveguides with corresponding dispersion profiles. The amount and variation range of a data set should be appropriately decided, otherwise it might waste time in generating data and the training network, or make the network fail to locate the intrinsic connection between the waveguide and its dispersion. Based on the data set, the network could be fed and trained.

In detail, given that it is almost impossible to obtain the analytical solution of the dispersion profiles, those in the data set were all numerical solutions. Each of the profiles contained many points, which are dispersion values corresponding to their wavelengths, and these points were calculated using Equation (1) mentioned above. The entire profile represents the dispersion variation of one set of structural parameters (six specific structural parameters form one set). For the essential data set, we use the six structural parameters [H1, H2, H3, Hs1, Hs2, W] and materials in simulation to obtain the dispersion values through indirect calculation from n_{eff}. Notably, we let H1 equal to H3, in order to reduce the computation in the research, as the change of H1 and H3 has a similar influence on waveguide's dispersion [6]. Therefore, the data contained dispersion values varying with wavelength, λ, which represent the dispersion profile mentioned above, as well as the corresponding set of structural parameters. The entire data set consisted of thousands of pieces of data.

The data set is divided into input and output parts. The input part contains the discretely dispersion profile [D1, D2, . . . , Dn], while the output part contained the structural parameters [H1, H2, H3, Hs1, Hs2, W]. The variation range of those parameters in the data set is shown in Table 1. Further, discrete dispersion values from 1500 nm to 3500 nm for As_2S_3 waveguide and 1000 nm to 3000 nm for Si3N4 waveguide with an interval of 50 nm were calculated. Specific wavelength bands were selected and some of the values taken out to consider the target in the training process. Prior to training, proper preprocessing and normalization for the original data is essential. Here, we simulated the structures again where the data are identified as null values due to the calculation error of software. Then, we converted all the structural parameters into range [0, 1] via linear mapping (for example, for parameter W, the maximum value in W is converted to 1 and the minimum one is converted to 0), in order to increase the calculating efficiency. The same conversion was also applied to the dispersion values where the interval was [−1, 1]. After that, the input and output data were fed into the network, which is trained by adjusting the weight of each layer and comparing the error value continuously. The error value, here, means the difference between the predicted values from network and the actual values in the data set. The training process stops after specific number of epochs in the network. One epoch means a circulation in network training, after which the error value will gradually decreases. The number of epochs selected in our model is 200, considering both the efficiency and accuracy. There will be a convergence of error value if proper settings of network are made. After completing a careful search and attempts were made to identify the settings corresponding to a much lower convergence (stated in Section 2.3), an inverse design model was used to obtain proper mapping between discrete dispersion and structural parameters.

Table 1. Structure parameter ranges of As2S3 and Si3N4 waveguides.

Material	Parameter	Min (nm)	Max (nm)
As_2S_3	W	880	975
	Hs1	97	122
	Hs2	102	125
	H1	310	340
	H2	945	975
	H3	310	340
Si_3N_4	W	1120	1280
	Hs1	155	180
	Hs2	145	170
	H1	300	325
	H2	950	975
	H3	300	325

2.3. Network for the Inverse Design

After data generation, the core of inverse design followed—the construction and training of the network. Establishing a network with an appropriate structure is essential, as it is determinant of prediction accuracy. In our work, Deep Neural Network (DNN) was utilized. Both input and output data can be regarded as one-dimensional column vectors, feeding into the DNN. The structure of DNN is shown in Figure 2b. An open-source artificial Neural Network library Keras, based on Python, is applied. In terms of the network, there are four hidden layers and 200 neurons in each layer. Linear rectification function Relu and Sigmoid are applied in four hidden layers and an output layer, respectively. We selected the mean square error (MSE) as the loss function and Adam algorithm as the optimizer. The loss function, here, demonstrates how we measure the error value between predicted value and actual value, and the optimizer is the method by which the network decreases the error level. We also use the dropout layer to weaken over fitting [30–32]. The detailed settings of network are shown in Table 2.

Table 2. Settings of the Neural Network.

Used NN Library	Hidden Layers	Neurons per Layer
Keras	4	200
Activation function	**Loss function**	**Optimizer**
Relu and Sigmoid	Mean Square Error	Adam algorithm

In order to confirm the performance and accuracy of the trained network, we compared and identified the error value between the predicted parameters obtained by network and the actual value. Utilizing the test data set, an objective dispersion value set [D1, D2, . . . , Dn] could be formulated and placed into the trained network, then structural parameters could be predicted [H1, H2, H3, Hs1, Hs2, W], the corresponding dispersion value could be calculated. The actual structures of that dispersion are already known from the test data. Understanding whether the error between the predicted and actual structures are low or not is standard for confirming the performance of the trained network. Given that our aim was to obtain waveguide structure where the dispersion characteristic satisfied the specific profiles, this straightforward factor represented the error of structural parameters between predicted and actual values. Considering this, a comparison of predicted and actual values is displayed in the following part of Sections 3.1 and 3.2.

For the inverse design process, a set of dispersion values (optimization aims) were used to meet the target performance, they were then fed into the trained network, in order to predict the corresponding structural parameters. Similarly, these predicted parameters were simulated to identify the real dispersion values, evaluate accuracy and the performance of the trained network. The procedure of inverse design is shown in the schematic of Figure 2b.

3. Results and Analysis

3.1. Inverse Design of As_2S_3 Waveguide

We generated the As_2S_3 waveguide data set with a scale of about 6000, and these data are divided into the train set and test set at a ratio of 80:20. Both the train and test set contain a number of complete pieces of data, which are used to train the network and check the performance of the network, respectively. Among the settings for the network, the ratio of 80:20 was one available option, and the algorithm automatically completes the division. The train set was used for training, including calculating the loss function and the optimizer to adjust the weight throughout the training process. Whereas, the test set was used for the network to check the values of loss function—it was not involved in the training process as it could only be used for testing. In this way, we ensured the reliability performance of the network evaluation via the test set. The results are displayed below.

For the inverse design of the As_2S_3 waveguide, we selected the target wavelength bands, ranging from 2000 nm to 2800 nm and from 1800 nm to 2900 nm, respectively (a wide and a narrow band), which means the input of network is on 9 [D1, D2, D3 ... , D9] dimensions or 11 [D1, D2, ... , D11] dimensions, and the output was made up of 6-dimensional structural parameters [H1, H2, H3, Hs1, Hs2, W].

For the different target wavelength bands, where the structures of the network are the same, the different combinations of dispersion values were extracted from the data set to feed the network as input data. This also represents the efficiency and reusability of inverse design method. We did not use the entire dispersion values (1500 nm to 3500 nm for As_2S_3 waveguide and 1000 nm to 3000 nm for Si_3N_4 waveguide) due to the influence of the intrinsic zero dispersion of materials, which result in a rapid variation of waveguide's dispersion inevitably. The reason for generating a data set with extra dispersion values was to retain the possibility of adjusting the target wavelength band freely by selecting different dispersion values as input of the network. This should not be time-consuming, and with different input data, the training process only lasts several tens to hundreds of seconds.

In Figure 3, the prediction of two of the six structural parameters were used to display the training results for examples. The target wavelengths, in this study, are 2000 nm to 2800 nm. The X-axis is the actual parameter, and the Y-axis is the predicted values obtained by setting target dispersion as the corresponding dispersion values from test set. This scatter diagram does not display all the testing results, which are in a large amount. As shown in Figure 3, most of points are close to the straight line of y = x with a small variation, which represents good performance of the network with the predicted values well close to actual values. We calculated the mean absolute error values of each structural parameters predicted by the trained network to check performance of the network more straightforward, as shown in Table 3. We found that the MAE values of those structural parameters are all less than 1 nm, which is highly within the tolerance for practical fabrication.

In order to achieve the practical application of photonic integrated circuit design, we set three goals for dispersion engineering: Broadband low dispersion in different wavelength bands, broadband constant dispersion with positive or negative values, and slope-maintained linear dispersion, where the target dispersion values are all 0, a certain constant value, and a linear function respectively. By applying the trained network, predicted structures corresponding to the target dispersion can be obtained. We place these structural parameters into simulation, calculate the actual dispersion profiles, then compare the actual and target dispersion for verification.

Table 3. Mean Absolute Error of predicted structural parameters.

Mean Absolute Error (MAE)	Hs1	Hs2	H1&H3	H2	W
As2S3 Slot Waveguide	0.24	0.16	0.38	0.45	0.55

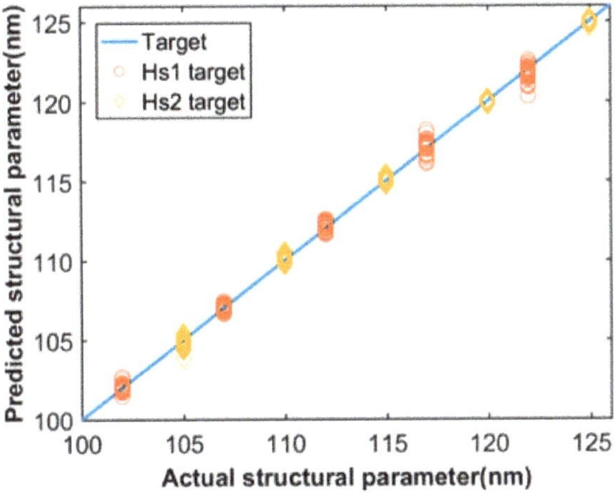

Figure 3. Actual and predicted data for As2S3 waveguide.

The results of the inversely designed dispersion profiles of As2S3 waveguide obtained from trained Neural Network are shown in Figure 4. The detailed structural parameters, mentioned in Figure 4, is shown in Table 4.

In Figure 4a, the target wavebands are 1800–2900 nm and 2000–2800 nm respectively, which represents a wider band and a narrower band. Horizontal dotted line in the expanded Figure refers to the target dispersion values, which remained at zero, and represents the broadband low dispersion. The corresponding structures are A1 and A2, predicted by the network. Dispersion of A1 varies between 0–4.5 ps/(nm·km) for a 1154-nm bandwidth from 1885 nm to 3039 nm and dispersion of A2 varies between 0–1.5 ps/(nm·km) for an 861-nm bandwidth from 1945 nm to 2806 nm. The results of the predicted structures do not exactly follow the target bands due to the error of network. We found that, given that Structure A1 has a longer target band, its flatness of dispersion curve is not better than that of A2. However, it has a wider wavelength range with a relatively low dispersion. Compared with Reference [6], a much lower dispersion in a slightly narrower bandwidth was obtained by the proposed network.

Table 4. The structural parameters of mentioned dispersion engineering waveguide.

Structure (nm)		Hs1	Hs2	H1&H3	H2	W
	A1	111.5	112	330	961	930
	A2	114.5	105	330	962	907.5
	A3	104	114	330	1010	911
	A4	101	116	329	919	880
	A5	98	114	323	1037	893
As2S3 Slot Waveguide	A6	107	116	329	892	898
	A7	103.5	102	328	960	923.5
	A8	111	101.5	325	958	915
	A9	100	123	330	957	880
	A10	98	126	330	961	847

Figure 4b,c show the results of broadband constant dispersion and slope maintained linear dispersion. The target waveband is 2000–2800 nm. Figure 4b shows five dispersion curves for structure A1, A3, A4, A5 and A6. The dispersion curve for A1 is for comparison. Broadband constant dispersion of 30, −30, 50, and −50 ps/(nm·km) can be predicted by the network. All five dispersion curves have wavebands of constant dispersion containing the target band, and the range of constant dispersion slightly exceeds the target band. As in Figure 4a, the horizontal dotted lines in Figure 4b refer to target values being fed into the network, corresponding to constant values mentioned above. Moreover, Figure 4c shows the linear dispersion curves with four different slopes, corresponding to structures A7, A8, A9, and A10 predicted by the network. The vertical dotted lines, lying on 2000 nm and 2800 nm, indicate the linear area, whereby the four dispersion slopes are maintained to be 0.04, 0.02, −0.02, −0.04 ps/(nm²·km), respectively. This kind of linear dispersion may be useful for compensation in fiber systems.

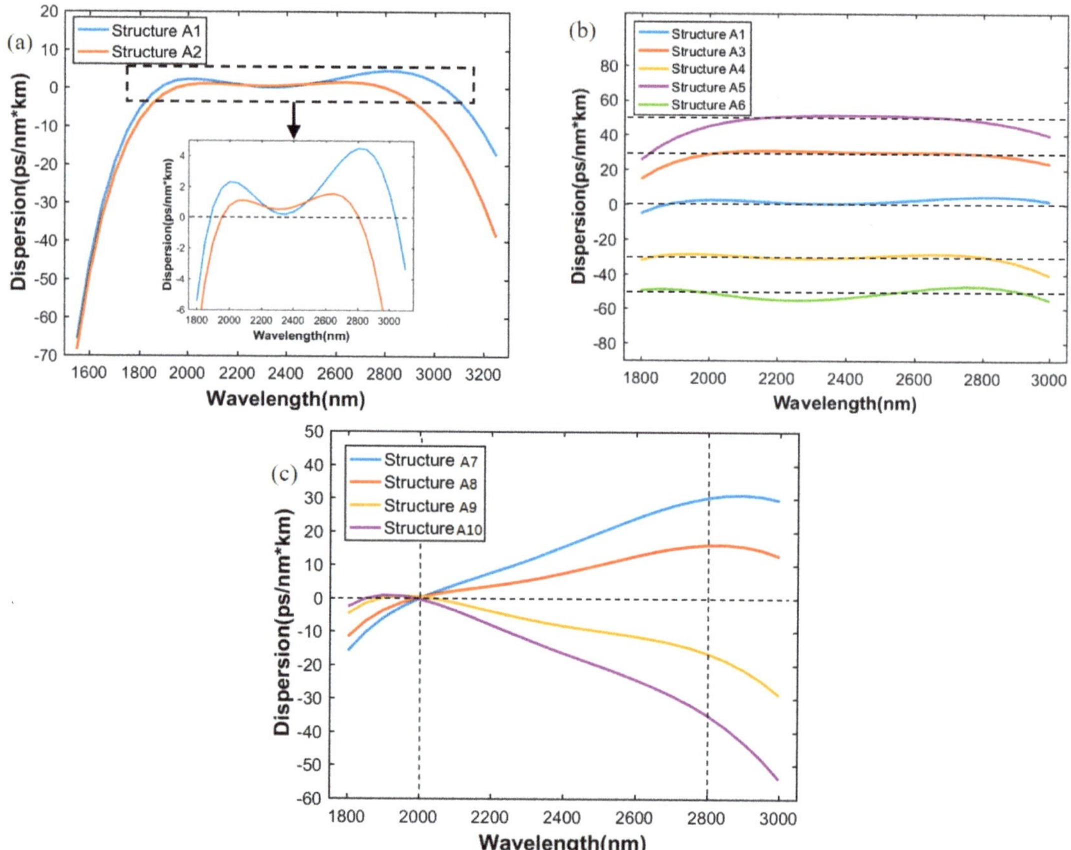

Figure 4. Inversely designed dispersion curves of As2S3 waveguides. (**a**) Broadband low dispersion for different target wavebands; (**b**) broadband maintained constant dispersion; (**c**) slope maintained linear dispersion.

3.2. Inverse Design of Si₃N₄ Waveguide

Our proposed dispersion engineering method can also be applied to Si3N4 double slot waveguide, and it proves universality in photonic waveguide design. The process of network training is basically similar to that of As₂S₃ waveguide, but the new data set needs to be generated for theSi₃N₄ waveguide. The selected target wavelength band range from 1300 nm to 2200 nm and 1200 nm to 2400 nm, a wider band and a narrower band. Different material dispersion and modal refractive index should be expected for Si₃N₄ waveguide.

The scatter diagram containing the prediction of two parameters for Si₃N₄ waveguide is shown in Figure 5, and the mean absolute error values of the results for Si₃N₄ waveguide are shown in Table 5. The target band is 1300 nm to 2200 nm. The MAE values are also at a lower level, indicating good performance of prediction. However, the performance of the inverse design for Si3N4 waveguide is not as good as that of As₂S₃ waveguide, in terms of the predict accuracy in scatter diagram and mean absolute error value in Table 5. In fact, there might be a best interval for scale of data set and variation range of the parameters in the data set, if deviated from this interval, the performance of network will decrease when other conditions remain unchanged. We attempted to find the best interval for Si_3N_4 waveguide, but a difference exists, maybe due to the sub-par interval for that of Si_3N_4 waveguide. Nonetheless, the results in Figure 6 show that the network for Si_3N_4 waveguide also possess satisfying performance, thus the difference between Tables 3 and 5 can be tolerated. The detailed structural parameters mentioned in Figure 6 are shown in Table 6.

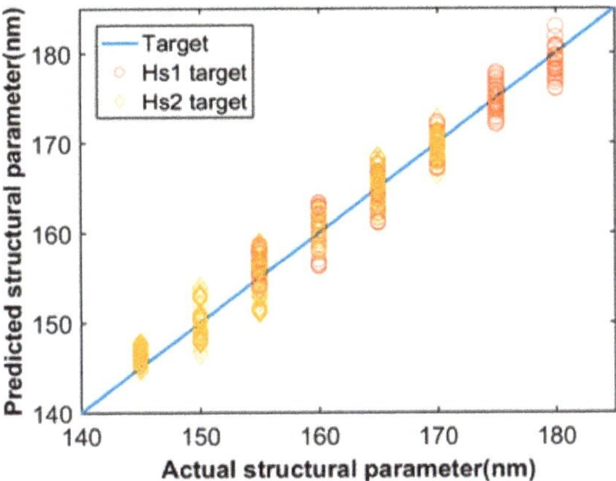

Figure 5. Actual and predicted data for Si3N4 waveguide.

Table 5. Mean Absolute Error of predicted structural parameters.

Mean Absolute Error (MAE)	Hs1	Hs2	H1&H3	H2	W
Si3N4 Slot Waveguide	1.27	1.88	1.13	0.99	2.33

As for the predicting results, in Figure 6a, for structure B1, the broadband low dispersion varies from 0 to 5.8 ps/(nm·km) ranging from 1226 nm to 2368 nm for a 1142-nm bandwidth and for structure B2, dispersion varies from 0 to 1.4 ps/(nm·km) from 1306 nm to 2067 nm for a 761-nm bandwidth. Horizontal dot lines in zoom-in figure are the target dispersion values, all zero here, which represents broadband low dispersion. These well performed results in terms of much broader and lower dispersion compared with Reference [7] would be highly useful for broadband nonlinear processing applications on Si3N4 platform.

Similar to As_2S_3, dispersion engineering, with maintained constant values or maintained dispersion slopes, is also carried out using the proposed method, as shown in Figure 6b,c. Constant dispersions of 30, −30, 50, and −50 ps/(nm·km), ranging from 1300 nm to 2200 nm with small variations, are obtained by structure B3 to B6. While, the horizontal dot lines are target values, corresponding to the certain dispersion. The dispersion slopes of 0.04, 0.02, −0.02, −0.04 ps/(nm²·km) from 1300 nm to 2200 nm are maintained, as indicated by the vertical dot lines, and obtained by Structure B7 to B10, as shown in Figure 6c.

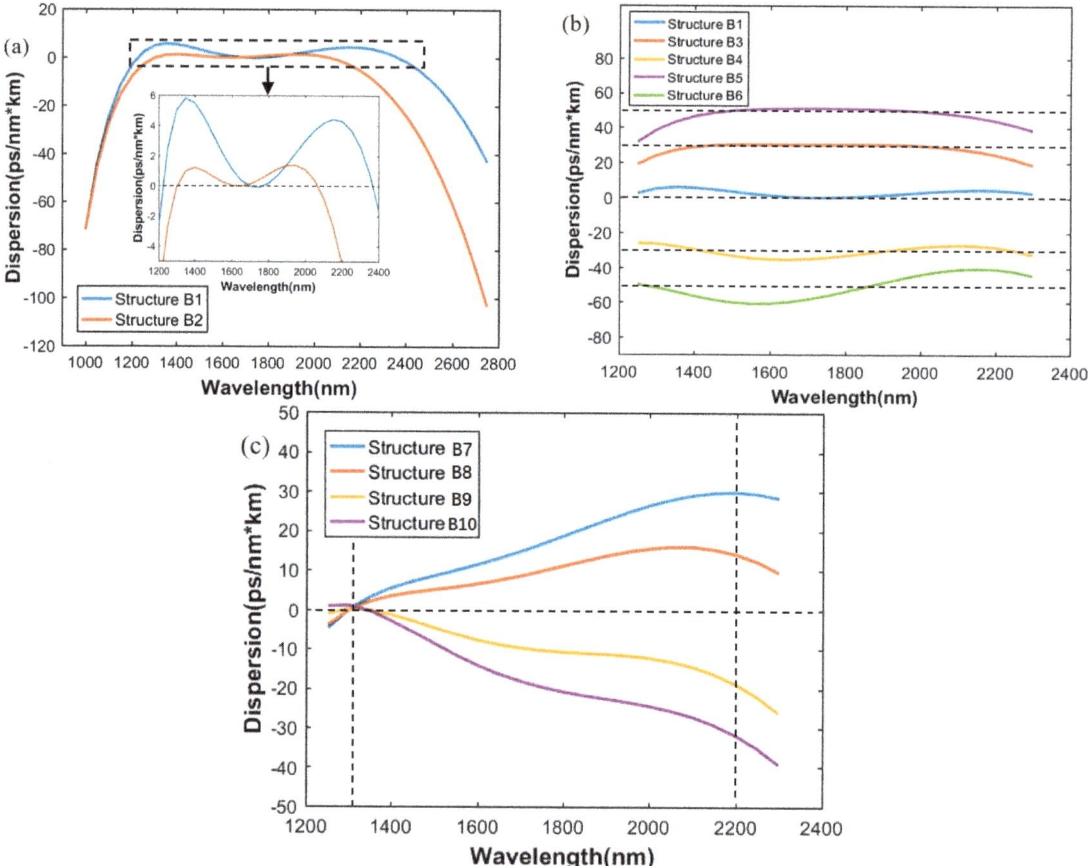

Figure 6. Inversely designed dispersion curves of Si3N4 waveguides. (**a**) Broadband low dispersion for different target wavebands; (**b**) broadband maintained constant dispersion; (**c**) slope maintained linear dispersion.

Table 6. The structural parameters of mentioned dispersion engineering waveguide.

Structure (nm)		Hs1	Hs2	H1&H3	H2	W
	B1	176.5	171.5	307	941	1309
	B2	161	149	318	939.5	1131
	B3	162	159.5	298	979	1216.5
	B4	161	172	327	889	1189
Si3N4 Slot Waveguide	B5	162	156	299	1035	1216.5
	B6	175	159	325	855	1189
	B7	165	147.5	307	953	1293
	B8	161	151	306.5	937	1218
	B9	168	156	324	938	1126
	B10	183	155	325	935	1120

3.3. Influence of Sidewall Angle in Fabrication

Considering the fabrication of the horizontal double slot waveguide, there always exists a sidewall angle α, which means the wall of the waveguide is not perfectly vertical, as shown in the Figure 7a. This may influence the light field distribution and then correspondingly change the waveguide dispersion. As shown in Figure 7b, the waveguide in the left is perfectly vertical and the waveguide on the right has a sidewall angel of 5 degrees.

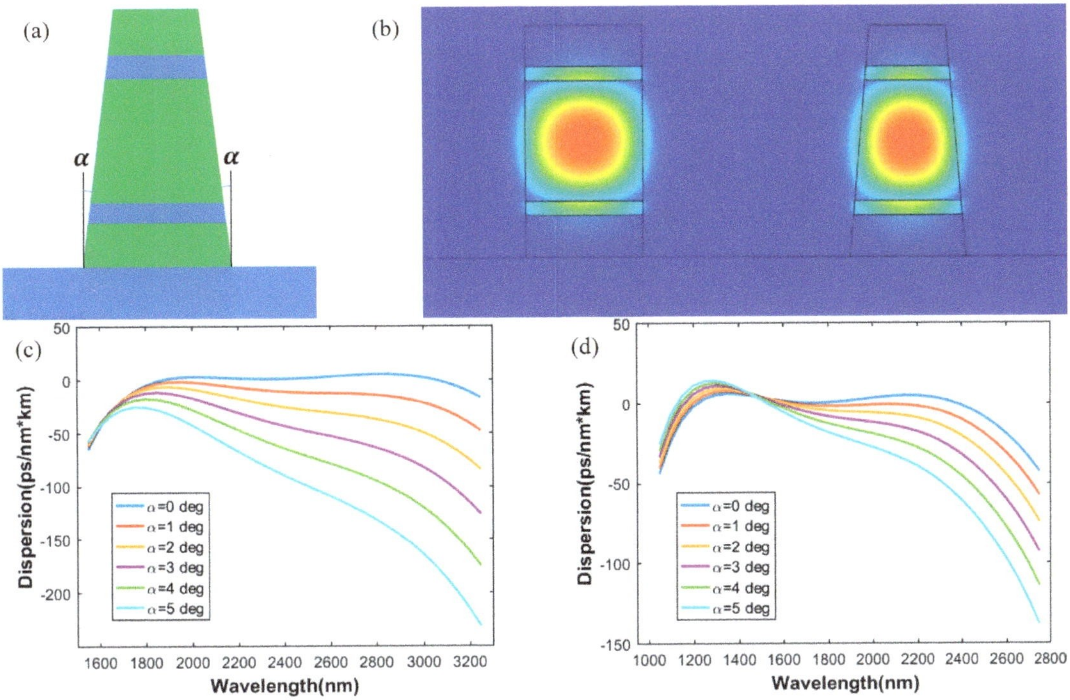

Figure 7. (**a**): Waveguide with sidewall angles; (**b**) fundamental modes for waveguides with sidewall angle of 0 and 5 degrees; (**c**) dispersion curves with different sidewall angles for Structure A1; (**d**) dispersion curves with different sidewall angles for Structure B1.

Using Structure A1 and B1 as examples, the dispersion of the waveguide, with different sidewall angles from 0 to 5 degrees, was investigated. The results are shown in Figure 7c,d, respectively. We found that when the sidewall angle increases, the dispersion curves of the waveguide bend slightly. The dispersion of As_2S_3 waveguide changes more significantly than that of Si_3N_4 waveguide. A relatively high precision for As_2S_3 and Si_3N_4 waveguide fabrication needs to be maintained to keep the sidewall angle at a small level, in order to obtain the designed dispersion profile.

The fabrication of horizontal slot waveguide has already been studied in several works [33,34]. In view of the silicon nitride horizontal slot waveguide, the thin film of silicon nitride and silica can be deposited on silica wafer in the specific order via low pressure or plasma enhanced chemical vapor deposition (LPCVD/PECVD). The depositing thickness of film can be controlled by the duration time of deposition. After that, a dry etching method, such as reactive ion etching (RIE), can be applied to complete the processing of waveguide, which will result in a high performance of fabrication.

3.4. Generation of Frequency Combs via Double Slot Micro-Ring Resonator

It can be inferred from the study in [29] that a well-designed flat and low anomalous dispersion is conducive for generating the Kerr frequency comb with board bandwidth and small power variation. We formulated micro-ring resonators with double-slot structure, with B1 and B2 specific structures. Then, we adopted their dispersion curve into a simulation of Kerr frequency comb, in order to verify the performance of practical application via inverse design process.

The dynamics of comb generation in micro-ring resonators is well-described by the Lugiato-Lefever equation, which can be described as Equation (2) below [35,36]:

$$\frac{\partial \widetilde{A}_\mu(t)}{\partial t} = \left(-\frac{\kappa}{2} + i(2\pi\delta_0) + iD_{\text{int}}(\mu)\right)\widetilde{A}_\mu$$

$$- ig\mathcal{F}[|A|^2 A]_\mu + \sqrt{\kappa_{ex}} \cdot s_{in} \tag{2}$$

where $\widetilde{A}_\mu$ and A means spectral and temporal envelopes of light field in resonators, respectively, and their relationship follows the Fourier transform. While, κ is loss rate of the resonator and κ_{ex} is coupling rate between bus waveguide and resonator, g is Kerr frequency shift, and $|s_{in}|^2$ represents the pump power. $D_{\text{int}}(\mu)$ means the integrated dispersion of μth frequency component and this can be calculated through the dispersion curve obtained above. δ_0 is pump detuning. In our simulation, FSR of comb is set to be 200 GHz, corresponding to L = 850 µm and R = 135 µm. The loss of waveguide (Si3N4) is set to be 0.1 dB/cm and the resonator is assumed critical coupling, which means $\kappa = \kappa_{ex}$. The Q value of resonator is 6.5×10^6. The power and wavelength of pump is 1 W and 1800 nm. The pump detuning is 56 times of resonator FWHM, around 4.88 GHz.

Based on the conditions mentioned above, the flat and broad spectrums of Kerr frequency comb are obtained from Structure B1 and B2, as shown in Figure 8, in order to compare influence of dispersion. The peak of spectrum in the longer wavelength area is generated because of the zero-value of integrated dispersion, according to Reference [9]. The spectrum in Figure 8a has a wide wavelength range and the power variation is relatively smooth. Quantificationally speaking, for a 10-dB power variation, the bandwidth of spectrum is measured to be 564 nm from 1515 nm to 2079 nm. For a 20-dB variation, the bandwidth of spectrum is measured to be 1068 nm from 1415 nm to 2483 nm. In Figure 8b, the spectrum of B2 possesses smaller power variation around the narrower band due to the smaller dispersion. For a 10-dB power variation, the band width of spectrum is measured to be 777 nm from 1416 nm to 2193 nm. It lacks bandwidth due to the narrower range of low dispersion. In addition, comb spectrum of strip waveguide is an envelope of square of hyperbolic secant [9]. In comparison, the spectrum of double slot waveguide has the potential to extend to low power variation. In practical terms, this could be adjusted according to the actual demands.

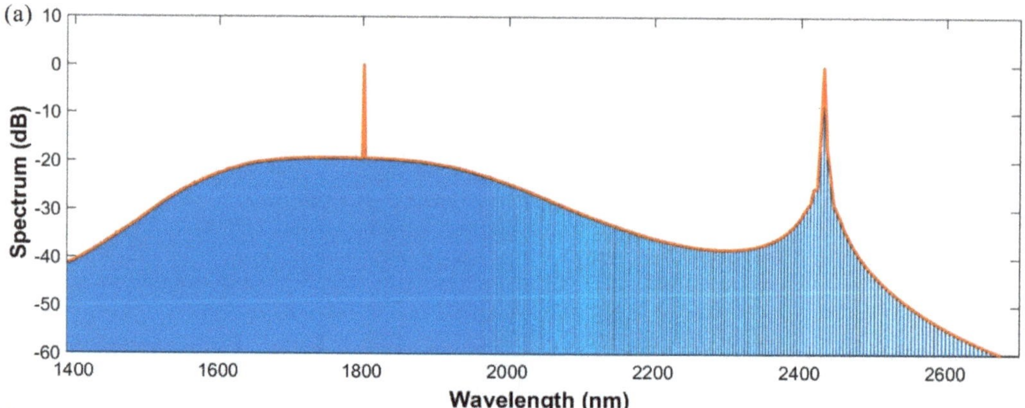

Figure 8. *Cont.*

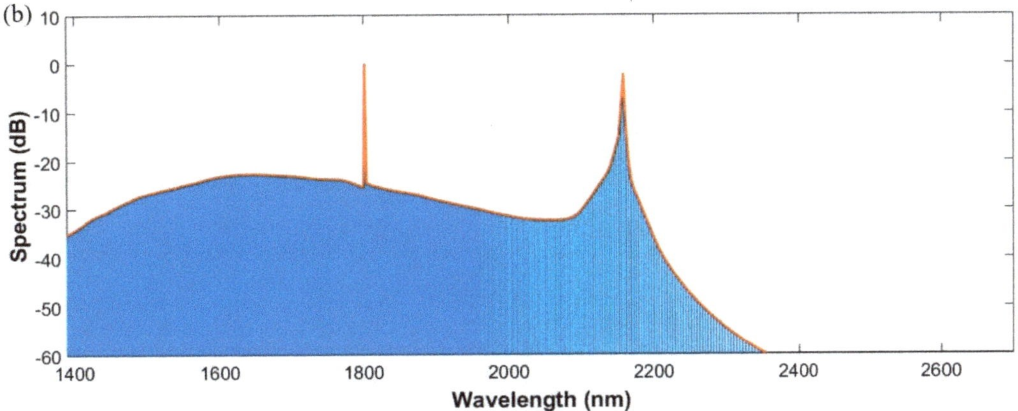

Figure 8. (**a**) Spectrum of Kerr frequency comb generated from micro-ring resonator with Structure B1; (**b**) spectrum of Kerr frequency comb generated from micro-ring resonator with Structure B2.

4. Conclusions

In this work, a Neural Network aided inverse design method is demonstrated for agile dispersion engineering of a horizontal double slot waveguide structure. The proposed objective-motivated, automation-supported inverse design method provides rapid optimization with high performance. Broadband low dispersion, constant dispersion and slope maintained linear dispersion were achieved for As_2S_3 and Si_3N_4 waveguides. These well-performed studies achieved results of versatile dispersion profiles that would be highly useful for broad applications. We believe this can be extended to very broad photonic devices to obtain ultimate performance far beyond dispersion. Significant potential can be expected for photonic design automation of integrated circuit.

Author Contributions: Data curation, Z.W. and J.L.; formal analysis, J.L.; methodology, Z.W., J.D. and W.S.; writing—original draft, Z.W. and W.S.; writing—review and editing, J.D. and Z.H. All authors have read and agreed to the published version of the manuscript.

Funding: This work is supported by National Key R&D Program of China (2018YFB1801804), National Natural Science Foundation of China (NSFC) (61935011, 61875124, 61675128).

Institutional Review Board Statement: Not applicable.

Informed Consent Statement: Not applicable.

Data Availability Statement: Not applicable.

Acknowledgments: Due to all the reviewers and editors for their careful work and pertinent suggestions.

Conflicts of Interest: The authors declare no conflict of interest.

References

1. Ramaswami, R.; Sivarajan, K.; Sasaki, G. *Optical Networks: A Practical Perspective*; Morgan Kaufmann: Burlington, MA, USA, 2009.
2. Gunning, F.C.G.; Corbett, B. Time to open the 2-μm window? *Opt. Photonics News* **2019**, *30*, 42–47. [CrossRef]
3. Haus, H.A.; Fujimoto, J.G.; Ippen, E.P. Structures for additive pulse mode locking. *J. Opt. Soc. Am. B* **1991**, *8*, 2068–2076. [CrossRef]
4. Madsen, C.; Lenz, G.; Bruce, A.; Cappuzzo, M.; Gomez, L.; Scotti, R. Integrated all-pass filters for tunable dispersion and dispersion slope compensation. *IEEE Photonics Technol. Lett.* **1999**, *11*, 1623–1625. [CrossRef]
5. Petrov, A.Y.; Eich, M. Dispersion compensation with photonic crystal line-defect waveguides. *IEEE J. Sel. Areas Commun.* **2005**, *23*, 1396–1401. [CrossRef]
6. Wang, Y.; Zhang, M.; Lu, L.; Li, M.; Wang, J.; Zhou, F.; Dai, J.; Deng, L.; Liu, D. Ultra-flat and broad optical frequency combs generation based on novel dispersion-flattened double-slot microring resonator. *Appl. Phys. A* **2016**, *122*, 11. [CrossRef]
7. Zhang, L.; Bao, C.; Singh, V.; Mu, J.; Yang, C.; Agarwal, A.M.; Kimerling, L.C.; Michel, J. Generation of two-cycle pulses and octave-spanning frequency combs in a dispersion-flattened micro-resonator. *Opt. Lett.* **2013**, *38*, 5122–5125. [CrossRef] [PubMed]

8. Kim, S.; Han, K.; Wang, C.; Jaramillo-Villegas, J.A.; Xue, X.; Bao, C.; Xuan, Y.; Leaird, D.E.; Weiner, A.M.; Qi, M. Dispersion engineering and frequency comb generation in thin silicon nitride concentric microresonators. *Nat. Commun.* **2017**, *8*, 1–8. [CrossRef]

9. Shun, F.; Tanabe, T. Dispersion engineering and measurement of whispering gallery mode microresonator for Kerr frequency comb generation. *Nanophotonics* **2020**, *9*, 1087–1104.

10. Xue, X.; Xuan, Y.; Liu, Y.; Wang, P.; Chen, S.; Wang, J.; Leaird, D.E.; Qi, M.; Weiner, A.M. Mode-locked dark pulse Kerr combs in normal dispersion microresonators. *Nat. Photonics.* **2015**, *9*, 594–600. [CrossRef]

11. Wang, C.; Bao, C.; Xuan, Y.; Han, K.; Leaird, D.E.; Qi, M.; Weiner, A.M. Normal Dispersion High Conversion Efficiency Kerr Comb with 50 GHz Repetition Rate. In Proceedings of the 2017 Conference on Lasers and Electro-Optics (CLEO), San Jose, CA, USA, 4–19 May 2017; p. SW4N5.

12. Matsko, A.; Savchenkov, A.A.; Maleki, L. Normal group-velocity dispersion Kerr frequency comb. *Opt. Lett.* **2012**, *37*, 43–45. [CrossRef]

13. Parra-Rivas, P.; Gomila, D.; Knobloch, E.; Coen, S.; Gelens, L. Origin and stability of dark pulse Kerr combs in normal dispersion resonators. *Opt. Lett.* **2016**, *41*, 2402–2405. [CrossRef] [PubMed]

14. Xu, W.; Zhou, L.; Lu, L.; Chen, J. Dispersion-engineered Optical Phased Array for Aliasing-free Beam-steering with a Plateau Envelope. In Proceedings of the 2018 Asia Communications and Photonics Conference (ACP), Hangzhou, China, 26–29 October 2018; pp. 1–3.

15. Yang, M.; Xu, L.; Wang, J.; Liu, H.; Zhou, X.; Li, G.; Zhang, L. An Octave-Spanning Optical Parametric Amplifier Based on a Low-Dispersion Silicon-Rich Nitride Waveguide. *IEEE J. Sel. Top. Quantum Electron.* **2018**, *24*, 1–7. [CrossRef]

16. Ji, X.; Yao, X.; Gan, Y.; Mohanty, A.; Tadayon, M.A.; Hendon, C.P.; Lipson, M. On-chip tunable photonic delayline. *APL Photonics* **2019**, *4*, 090803. [CrossRef]

17. Dudley, J.M.; Coen, S. Coherence properties of supercontinuum spectra generated in photonic crystal and tapered optical fibers. *Opt. Lett.* **2002**, *27*, 1180–1182. [CrossRef]

18. Hui, Z.; Yang, M.; Pan, D.; Zhang, T.; Gong, J.; Zhang, M.; Zeng, X. Slot–slot waveguide with negative large and flat dispersion covering C+L+U waveband for on-chip photonic networks. *Appl. Opt.* **2019**, *58*, 5728–5739. [CrossRef]

19. Okawachi, Y.; Saha, K.; Levy, J.S.; Wen, Y.; Lipson, M.; Gaeta, A.L. Octave-spanning frequency comb generation in a silicon nitride chip. *Opt. Lett.* **2011**, *36*, 3398–3400. [CrossRef]

20. Saha, K.; Okawachi, Y.; Levy, J.S.; Lau, R.K.W.; Luke, K.; Foster, M.A.; Lipson, M.; Gaeta, A.L. Broadband parametric frequency comb generation with a 1-mm pump source. *Opt. Express* **2012**, *10*, 26935–26941. [CrossRef]

21. He, Z.; Du, J.; Chen, X.; Shen, W.; Huang, Y.; Wang, C.; Xu, K.; He, Z. Machine learning aided inverse design for few-mode fiber weak-coupling optimization. *Opt. Express* **2020**, *28*, 21668–21681. [CrossRef]

22. Hao, J.; Zheng, L.; Yang, D.; Guo, Y. Inverse Design of Photonic Crystal Nanobeam Cavity Structure via Deep Neural Network. In Proceedings of the 2019 Asia Communications and Photonics Conference (ACP), Chengdu, China, 2–5 November 2019; p. M4A.296.

23. Unni, R.; Yao, K.; Zheng, Y. Deep Convolutional Neural Network for the Inverse Design of Layered Photonic Structures. In Proceedings of the 2020 Conference on Lasers and Electro-Optics (CLEO), San Jose, CA, USA, 10–15 May 2020; p. JW2D.14.

24. Chen, Y.; Du, J.; Huang, Y.; Xu, K.; He, Z. Intelligent gain flattening in wavelength and space domain for FMF Raman amplification by machine learning based inverse design. *Opt. Express* **2020**, *28*, 11911–11920. [CrossRef]

25. Shen, W.; Zeng, P.; Yang, Z.; Xia, D.; Du, J.; Zhang, B.; Xu, K.; He, Z.; Li, Z. Chalcogenide glass photonic integration for improved 2 μm optical interconnection. *Photonics Res.* **2020**, *8*, 1484–1490. [CrossRef]

26. Almeida, V.R.; Xu, Q.; Barrios, C.A.; Lipson, M. Guiding and confining light in void nanostructure. *Opt. Lett.* **2004**, *29*, 1209–1211. [CrossRef]

27. Xu, L.; Ni, X.; Liu, B.; Li, Y.; Hu, M. Ultraflat and low dispersion in a horizontal silicon nitride slot waveguide at near-infrared wavelengths. *Opt. Eng.* **2016**, *55*, 037109. [CrossRef]

28. Bao, C.; Yan, Y.; Zhang, L.; Yue, Y.; Ahmed, N.; Agarwal, A.M.; Kimerling, L.C.; Michel, J.; Willner, A.E. Increased bandwidth with flattened and low dispersion in a horizontal double-slot silicon waveguide. *J. Opt. Soc. Am. B* **2015**, *32*, 26–30. [CrossRef]

29. Weber, M.J. *Handbook of Optical Material*; CRC Press: Boca Raton, FL, USA, 2002.

30. Kingma, D.P.; Ba, J. Adam: A method for stochastic optimization. *arXiv* **2014**, arXiv:1412.6980.

31. Srivastava, N.; Hinton, G.; Krizhevsky, A.; Sutskever, I.; Salakhutdinov, R. Dropout: A simple way to prevent neural networks from overfitting. *J. Mach. Learn. Res.* **2014**, *15*, 1929–1958.

32. Goodfellow, I.; Bengio, Y.; Courville, A. *Deep Learning: Adaptive Computation and Machine Learning*; MIT Press: Cambridge, MA, USA, 2016.

33. Hayama, Y.; Uchibori, S.; Nakatsuhara, K. Fabrication of Nb_2O_5 horizontal slot waveguide structures. In Proceedings of the 2019 24th OptoElectronics and Communications Conference (OECC) and 2019 International Conference on Photonics in Switching and Computing (PSC), Fukuoka, Japan, 7–11 July 2019; pp. 1–3.

34. Liu, Y.; Chang, L.; Li, Z.; Liu, L.; Guan, H.; Li, Z. Polarization beam splitter based on a silicon nitride–silica–silicon horizontal slot waveguide. *Opt. Lett.* **2019**, *44*, 1335–1338. [CrossRef] [PubMed]

35. Lugiato, L.A.; Lefever, R. Spatial Dissipative Structures in Passive Optical Systems. *Phys. Rev. Lett.* **1987**, *58*, 2209. [CrossRef] [PubMed]

36. Guo, H.; Lucas, E.; Pfeiffer, M.H.; Karpov, M.; Anderson, M.; Liu, J.; Geiselmann, M.; Jost, J.D.; Kippenberg, T.J. Intermode Breather Solitons in Optical Micro-resonators. *Phys. Rev. X* **2017**, *7*, 041055.

Communication

Raman Amplification Optimization in Short-Reach High Data Rate Coherent Transmission Systems †

Mingming Tan [1,*], Md Asif Iqbal [2], Tu T. Nguyen [1,3], Paweł Rosa [4], Lukasz Krzczanowicz [1,5], Ian. D. Phillips [1], Paul Harper [1] and Wladek Forysiak [1]

[1]   Aston Institute of Photonic Technologies, Aston University, Birmingham B4 7ET, UK; nguyentu.hcmuns@gmail.com (T.T.N.); lukkr@fotonik.dtu.dk (L.K.); i.phillips@aston.ac.uk (I.D.P.); p.harper@aston.ac.uk (P.H.); W.Forysiak@aston.ac.uk (W.F.)
[2]   BT Applied Research, Adastral Park, Ipswich IP5 3RE, UK; mdasif.iqbal@bt.com
[3]   Infinera Pennsylvania, 7360 Windsor Dr, Allentown, PA 18106, USA
[4]   National Institute of Telecommunications, 04-894 Warsaw, Poland; P.Rosa@il-pib.pl
[5]   Department of Photonics Engineering, Technical University of Denmark, 2800 Kgs. Lyngby, Denmark
*   Correspondence: m.tan1@aston.ac.uk
†   This manuscript is extension version of the conference paper: Tan, M.; Iqbal, M.A.; Krzczanowicz, L.; Phillips, I.D.; Harper, P.; Forysiak, W. Optimization of Raman Amplification Schemes for Single-Span High Data Rate Coherent Transmission Systems. In Proceedings of the CLEO: Science and Innovations 2021, San Jose, CA, USA, 9–14 May 2021.

Citation: Tan, M.; Iqbal, M.A.; Nguyen, T.T.; Rosa, P.; Krzczanowicz, L.; Phillips, I.D.; Harper, P.; Forysiak, W. Raman Amplification Optimization in Short-Reach High Data Rate Coherent Transmission Systems. *Sensors* **2021**, *21*, 6521. https://doi.org/10.3390/s21196521

Academic Editors: Yang Yue, Jian Zhao, Jiangbing Du and Yange Liu

Received: 26 August 2021
Accepted: 23 September 2021
Published: 29 September 2021

Publisher's Note: MDPI stays neutral with regard to jurisdictional claims in published maps and institutional affiliations.

Abstract: We compared the transmission performances of 600 Gbit/s PM-64QAM WDM signals over 75.6 km of single-mode fibre (SMF) using EDFA, discrete Raman, hybrid Raman/EDFA, and first-order or second-order (dual-order) distributed Raman amplifiers. Our numerical simulations and experimental results showed that the simple first-order distributed Raman scheme with backward pumping delivered the best transmission performance among all the schemes, notably better than the expected second-order Raman scheme, which gave a flatter signal power variation along the fibre. Using the first-order backward Raman pumping scheme demonstrated a better balance between the ASE noise and fibre nonlinearity and gave an optimal transmission performance over a relatively short distance of 75 km SMF.

Keywords: Raman amplification; optical fibre communication

1. Introduction

In unrepeatered coherent transmission systems, distributed Raman amplification (DRA) can provide a better signal-to-noise ratio (SNR) than lumped (EDFA, discrete Raman amplification) or hybrid amplification techniques (hybrid Raman/EDFA) [1–7]. Particularly, higher-order DRA (second-order) has been advantageous in the transmission performance compared with first-order DRA because it minimises the signal power variation along the fibre and demonstrates a better balance between the amplified spontaneous emission (ASE) noise and the Kerr nonlinearities of the optical fibre in long-haul transmission systems [4]. However, as the data capacity of the optical transceiver has been increased from 100 Gb/s PM-QPSK to 600 Gb/s PM-64QAM, the maximum reach has significantly decreased from several thousand kilometres with QPSK to metro-network or data-centre-interconnect (DCI) with 64QAM [8,9], which can usually go up to a hundred kilometres or more. In such applications, the optimisation of the amplification technique remains to be investigated, so we would like to determine which amplification scheme delivers superior transmission performance over a relatively short fibre length with dual-polarisation 69 GBaud 64QAM signals.

In this paper, we expand our work on the optimisation of amplifiers in [10] and numerically and experimentally evaluate the performances of different representative amplification techniques used for transmission over 75.6 km SMF. The following discrete, hybrid, and distributed optical amplifiers are considered: EDFA, discrete Raman amplification, hybrid Raman/EDFA, first-order Raman-only amplification, and second-order (dual-order) distributed Raman amplification. First, we characterise the signal and noise power profiles of each scheme numerically and experimentally, and then we conduct the measurement of optical signal-to-noise ratio (OSNR) with 1 channel and 11 channels. It is shown that the noise level of the second-order DRA is indeed slightly lower than that in the other schemes. Finally, we experimentally evaluate how optical amplifiers perform using an 11-channel WDM grid with a 600 Gbit/s (69.4 Gbaud, 832 Gbit/s line rate) PM-64QAM real-time transceiver over 75 km SMF. Although our OSNR characterisations show that the second-order distributed Raman amplification had the lowest ASE noise level, the first-order distributed Raman amplification gave the best transmission performance, demonstrating the optimum balance between the linear noise and the fibre nonlinearities. Based on the simulated signal and noise power profiles, our transmission simulations show the same results as the experiment. For a short-reach metro network or DCI application with high-data-rate transceivers, the distributed Raman amplifier delivered the best transmission performance, compared with any other amplification scheme, including hybrid Raman/EDFA, discrete Raman, and EDFA only. However, the first-order distributed Raman scheme, with a simpler setup and lower pump power, performed better than the second-order scheme.

2. Experimental Setup and Characterization of All Amplification Schemes

Figure 1 shows our experimental setup. We conducted an unrepeated single-span experiment to evaluate the transmission performances of different optical amplification schemes using a high-data-rate signal. A commercially available transceiver was used to generate a 69.4 Gbaud polarization-division-multiplexed (PDM)-64QAM signal (50 Gbaud for the signal and the remainder for the FEC and other overheads), or a 34.7 Gbaud PDM-64QAM signal, corresponding to 600 Gbit/s and 300 Gbit/s data capacity, respectively. Ten channelized ASE signals spaced at 100 GHz (ranging from 1543.73 nm to 1551.72 nm) and the PDM-64QAM modulated signal centred at 1547.72 nm were combined via a 95/5 coupler to give an 11-channel WDM grid. The output signal was amplified by an EDFA and attenuated by a variable optical attenuator (VOA) to adjust the signal launch power to the fibre span. The amplified fibre link was a 75.6 km SMF with a WDM coupler (~0.8 dB attenuation) which gave a total attenuation of ~15.3 dB. The output signal from the link was amplified by an EDFA and then attenuated by a VOA in order to maintain a constant input power of −5 dBm to the receiver. The built-in real-time DSP algorithm was used to compensate for the linear impairments, and the BERs were calculated over 1 trillion (10^12) bits.

There were five amplification schemes investigated over the single 75.6 km SMF span, as illustrated on the right side of Figure 1, and the experimentally measured and simulated signal power profiles along the fibre are shown in Figure 2a. Scheme (a) was EDFA only to compensate the overall ~15 dB loss from the fibre and the WDM coupler, where the signal power in dBm decreased linearly along the fibre as shown in Figure 2a and amplified within the last few metres of the fibre. Scheme (b) was a discrete Raman amplifier that used two first-order Raman pumps at 1425 nm (~320 mW pump power) and 1445 nm (~350 mW pump power). Similarly, this is also a lumped amplification scheme, but 7.5 km of inverse dispersion fibre (IDF) was used as the Raman gain medium [5]. The main drawback of discrete Raman amplification in scheme (b) was the relatively long length of the gain fibre (7.5 km in this case), causing higher loss and higher accumulated fibre nonlinearity, compared to only a few metres in the EDFA case. Figure 2b clearly illustrates that the signal power using the discrete Raman amplifier decreased linearly, as was observed with the EDFA, but the amplification started within the last ~8 km. This extra length of Raman gain fibre could potentially increase both the linear ASE noise and the fibre nonlinearity when conducting the signal transmission. Scheme (c) was a hybrid amplification scheme combining the DRA and EDFA, which used first-order backward

(BW) Raman pumping and an EDFA [6,9]. The first-order DRA provided ~5 dB Raman gain with 160 mW pump power at 1455 nm, and the EDFA provided ~10 dB gain. Thus, as demonstrated in Figure 2a, the signal power variation along the fibre was ~10 dB compared with 15 dB using the EDFA-only or discrete Raman scheme. Scheme (d) was a distributed Raman amplifier using a first-order BW-propagated Raman pump at 1455 nm. The pump power at 1455 nm was ~410 mW, giving a signal power variation of ~6.5 dB. Scheme (e) was effectively a dual-order BW Raman pumping scheme, which used a second-order 1365 nm pump in addition to the 1455 nm seed pump. The pump power was ~940 mW at 1365 nm and ~25 mW at 1455 nm, giving a total pump power that was significantly higher than that in the other schemes but giving only ~4 dB signal power variation, which was the lowest of all the amplification schemes tested [2,3,10,11].

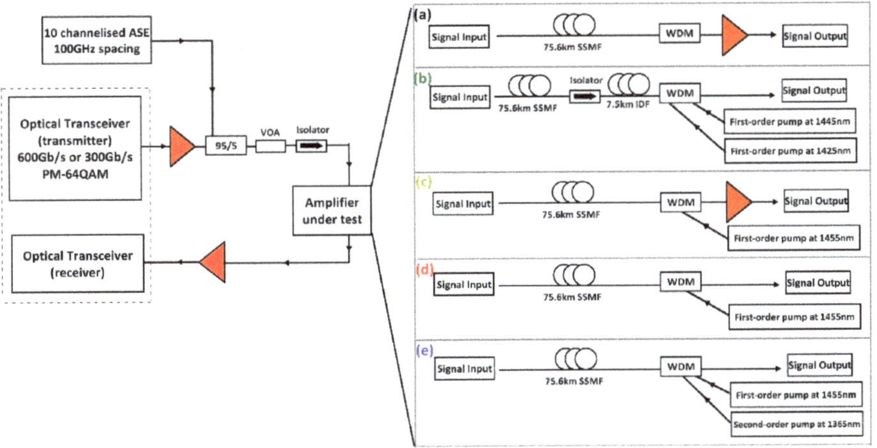

Figure 1. Experimental setup showing high-data-rate transmission with different amplification schemes: (**a**) EDFA only; (**b**) discrete Raman amplifier; (**c**) hybrid Raman/EDFA; (**d**) first-order distributed Raman amplifier; (**e**) second-order distributed Raman amplifier.

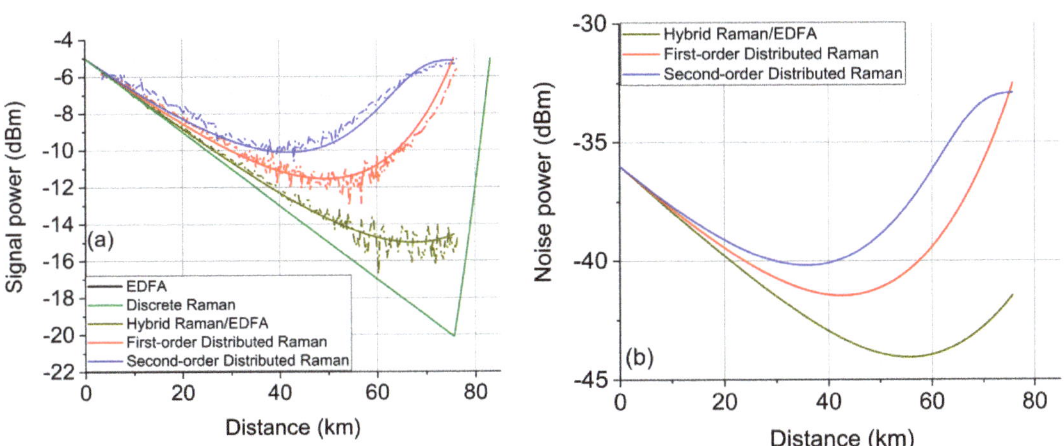

Figure 2. (**a**) Experimentally measured (dotted line) and simulated (solid line) signal power profiles along the fibre with five different amplification schemes; (**b**) Simulated noise power profiles along the fibre.

Figure 2b shows the simulated noise power profiles along the fibre for schemes (c), (d), and (e) from Figure 1. In EDFA and discrete Raman schemes (Figure 1a,b), the fixed ASE noise was added at the end of the fibre. In the discrete Raman amplifier, the noise was generated and accumulated over the last 8 km of the gain fibre, resulting in the highest noise power of all schemes—even higher than the EDFA scheme [12]. The hybrid Raman/EDFA was partially a distributed Raman amplifier where the noise was distributed over the whole transmission fibre, with the exception of EDFA noise which was added at the very last point of the fibre, increasing the overall noise level. From the two distributed schemes, the noise level of the second-order scheme was higher than that of the first-order scheme for the first ~73 km; however, within the last 2 km, the noise power of the first-order scheme dramatically exceeded the noise of the second-order scheme.

Figure 3 compares WDM (a) and a single channel (b) output signal spectra for all amplification schemes after the 75.6 km transmission span. The two DRA schemes showed significantly lower ASE noise levels (potentially higher signal-to-noise ratio) compared with the other three schemes, by approximately 5–6 dB, which means that the transmission performance in the linear regime is likely to show the same trend. The second-order distributed Raman scheme (blue) performed the best in terms of received OSNR, followed by the first-order scheme (red) with a slightly higher noise level as numerically shown in Figure 2b. This means that the superior noise performance using the second-order pumping was not obvious in the single 75.6 km span.

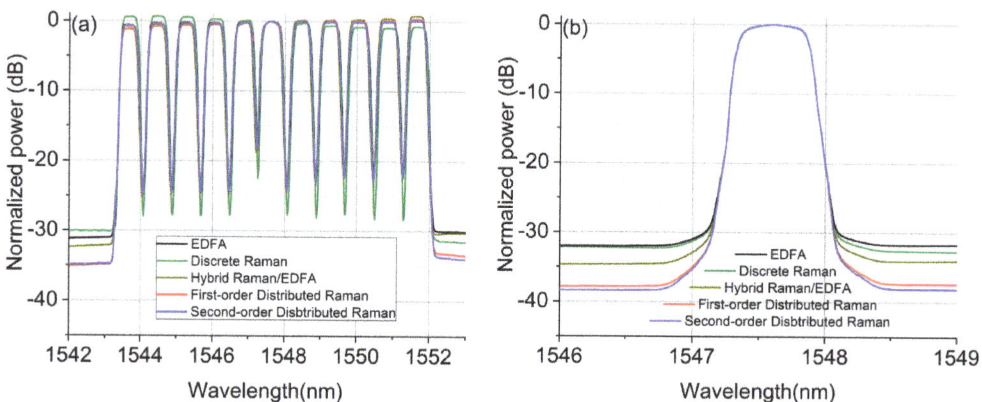

Figure 3. The output spectra from different optical amplifiers over a 75.6 km SMF: (**a**) with 11-channel input; (**b**) with 1 channel input.

3. Results and Discussion

The transmission performance was tested experimentally using a PM-64QAM WDM signal centred at 1547.72 nm and confirmed through simulation with signals of 600 Gb/s (Figure 4a) and 300 Gb/s (Figure 4b). We conducted a numerical simulation of the transmission performance of the PM-64QAM system, taking into account the simulated signal and noise power profiles for each amplification scheme. The simulation setup was similar, a PRBS length of 2^{16}-1 was used instead of 2^{15}-1. The propagation of the dual-polarisation complex envelope of an optical signal in optical fibre is governed by the coupled nonlinear Schrödinger equations (Manakov equations) and was simulated using the well-known split-step Fourier method [13,14], with a step size of 0.3 km using the simulated signal power profiles shown in Figure 2a. The noise from EDFAs implemented in the experiments at the transmitter and the receiver was taken into account in the simulations (ASE noise power density for each EDFA was approximately −155 dBm/Hz). To emulate EDFA noise, Gaussian noise with PSD of −144 dBm/Hz and −149 dBm/Hz was added at the end of the fibre for EDFA-only and hybrid Raman/EDFA, respectively. In addition, for the two distributed schemes, Raman noise was simulated as Gaussian noise, which was added to the signal after each step (0.3 km), aligning with the simulated noise profiles shown in Figure 2b. We used the same signal and noise power profiles for all the signal launch powers for simplicity. At the receiver, after coherent detection, the channel under test was filtered using an ideal low-pass filter. In order to take the imperfection of the DSP chain used in the experiments into account, the simulation results were also normalized by the same amount of the experimentally maximum achievable SNR, which was fixed at 20 dB.

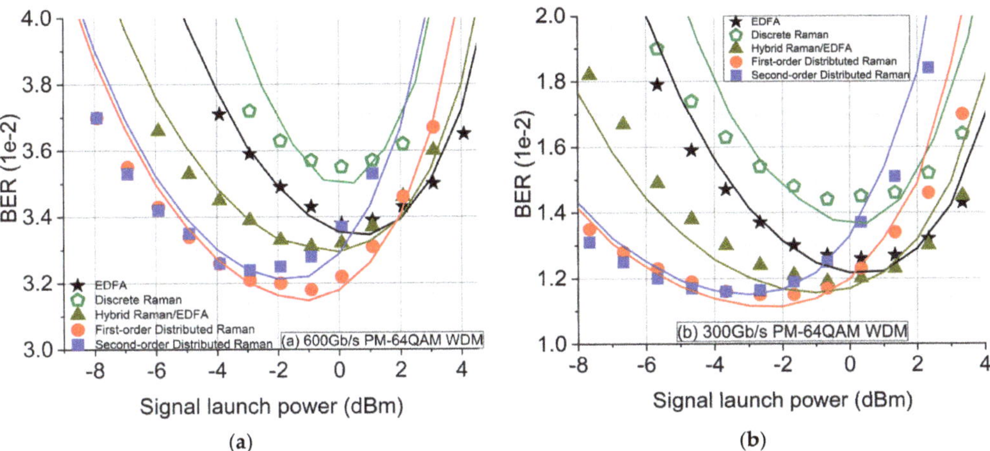

Figure 4. Simulated and experimental BER performances over a 75.6 km SMF with different amplification schemes measured at the centred channel (at 1547.72 nm): (**a**) 600 Gb/s PM-64QAM WDM transceiver; (**b**) 300 Gb/s PM-64QAM WDM transceiver.

Figure 4a,b shows the BERs versus signal launch power for 600 Gbit/s and 300 Gbit/s 64QAM signals. The solid lines are the numerical simulation results, and the dots are experimental results. The BERs at the optimum powers for EDFA, discrete Raman, and hybrid Raman/EDFA agree with the noise level analysis above; of these three schemes which used lumped amplification, the hybrid scheme showed the best BER as being more distributed, followed by the EDFA, and the discrete Raman scheme gave the worst BER performance due to the long Raman gain fibre at the end of the span. All three of these schemes performed significantly worse than the two distributed Raman schemes. The first-order Raman scheme showed better transmission performances at the optimum signal launch power for both signals and data rates. In the linear regime, the BERs with the second-order Raman scheme almost overlapped with the first-order Raman scheme. This means that the benefit from using the higher-order Raman pumping was not revealed in such a short transmission distance because the transmission performance in the linear regime was also influenced by the noise from the transceiver [7] and the EDFAs at the transmitter and the receiver. However, in the nonlinear regime, the first-order scheme gave a larger signal power variation along the fibre, as shown in Figure 3a, and consequently had a lower average signal power. Therefore, the first-order DRA scheme showed significantly greater tolerance against the fibre nonlinearity and better transmission performances than the second-order DRA. The experimental results were confirmed with our numerical simulations, as illustrated in Figure 4. As for short-reach high-data-rate coherent transmission systems, the simple first-order distributed Raman scheme, requiring low pump power, gave the best transmission performance compared with any other scheme, including the dual-order Raman scheme, hybrid Raman/EDFA, discrete Raman, and EDFA-only schemes.

4. Conclusions

In this paper, we demonstrate the experimental and numerical characterisation and optimisation for representative optical amplifiers, including an EDFA, a discrete Raman amplifier, a hybrid Raman/EDFA, first-order only, and second-order (dual-order) distributed Raman amplifiers, with a 600 Gb/s PM-64QAM transceiver (11-channel WDM grid) over a 75.6 km SMF. Our stand-alone characterisation results demonstrate that the second-order Raman scheme had flatter signal power profiles along the fibre, the lowest ASE noise level, and the highest OSNR. However, in the experimental transmission test, the first-order distributed Raman amplifier gave the best overall transmission overall performance. In the linear regime, the improvement introduced by higher-order pumping was not apparent, and therefore the first-order scheme showed similar performance to the second-order Raman scheme. However, because of the lower average signal power, the first-order scheme showed significantly superior transmission performance in the nonlinear regime in comparison with the second-order scheme. Therefore, the simpler first-order scheme gave the optimum balance between the linear noise and fibre nonlinearities in a single-span system with a high-data-rate transceiver. In addition, both distributed schemes demonstrated better BERs than the hybrid and discrete schemes. As expected, the hybrid Raman/EDFA scheme showed better performance than discrete schemes. Due to the extra 7.5 km Raman gain fibre, the discrete Raman scheme performed worst among all the amplification schemes considered.

Author Contributions: M.T. and M.A.I. proposed the concept and initiated the study; M.T., P.R., T.T.N. and M.A.I. carried out numerical simulations; M.T., M.A.I. and L.K. performed the experiments; M.T., M.A.I., P.R. and T.T.N. conducted analytical calculations; I.D.P., P.H. and W.F. supervised the studies. The paper was written by all authors. All authors have read and agreed to the published version of the manuscript.

Funding: UK Engineering and Physical Sciences Research Council (EPSRC) Grant EP/S003436/1 (PHOS), EP/V000969/1 (ARGON), Polish Ministry of Science and Higher Education Grant 12300060, and National Natural Science Foundation of China Grant 61975027.

Institutional Review Board Statement: Not applicable.

Informed Consent Statement: Not applicable.

Data Availability Statement: Not applicable.

Acknowledgments: We acknowledge industrial support from Lumentum (UK), II-VI, and Sterlite. The original data for this work is available at https://doi.org/10.17036/researchdata.aston.ac.uk.00 000522. We would like to thank Son Thai Le from Nokia Bell Lab, US, for their insightful discussions.

Conflicts of Interest: The authors declare no conflict of interest.

References

1. Ania-Castañón, J. Quasi-lossless transmission using second-order Raman amplification and fibre Bragg gratings. *Opt. Express* **2004**, *12*, 4372–4377. [CrossRef] [PubMed]
2. Bouteiller, J.-C.; Brar, K.; Bromage, J.; Radic, S.; Headley, C. Dual-order Raman pump. *IEEE Photonics Technol. Lett.* **2003**, *15*, 212–214. [CrossRef]
3. Jia, X.H.; Rao, Y.J.; Peng, F.; Wang, Z.N.; Zhang, W.L.; Wu, H.J.; Jiang, Y. Random-lasing-based distributed fiber-optic amplification. *Opt. Express* **2013**, *21*, 6572–6577. [CrossRef] [PubMed]
4. Tan, M.; Rosa, P.; Le, S.T.; Iqbal, M.A.; Phillips, I.D.; Harper, P. Transmission performance improvement using random DFB laser based Raman amplification and bidirectional second-order pumping. *Opt. Express* **2016**, *24*, 2215–2221. [CrossRef] [PubMed]
5. Rosa, P.; Tan, M.; Le, S.T.; Phillips, I.D.; Ania-Castañón, J.; Sygletos, S.; Harper, P. Unrepeatered DP-QPSK Transmission Over 352.8 km SMF Using Random DFB Fiber Laser Amplification. *IEEE Photonics Technol. Lett.* **2015**, *27*, 1189–1192. [CrossRef]
6. Lundberg, L.; Andrekson, P.; Karlsson, M. Power consumption analysis of Hybrid EDFA/Raman amplifiers in long-haul transmission systems. *J. Lightwave Technol.* **2017**, *35*, 2132–2142. [CrossRef]
7. Krzczanowicz, L.; Iqbal, M.A.; Phillips, I.D.; Tan, M.; Skvortcov, P.; Harper, P.; Forysiak, W. Low transmission penalty dual-stage broadband discrete Raman amplifier. *Opt. Express* **2018**, *26*, 7091–7097. [CrossRef] [PubMed]

8. Zhang, H.; Zhu, B.; Pfau, T.; Aydinlik, M.; Nadarajah, N.; Park, S.; Geyer, J.C.; Chen, L.; Aroca, R.; Doerr, C.; et al. Real-time transmission of single-carrier 400 Gb/s and 600 Gb/s 64QAM over 200km-span link. In Proceedings of the 45th European Conference on Optical Communication (ECOC 2019), Dublin, Ireland, 22–26 September 2019.
9. Galdino, L.; Lavery, D.; Liu, Z.; Balakier, K.; Sillekens, E.; Elson, D.; Saavedra, G.; Killey, R.I.; Bayvel, P. The Trade-off between Transceiver Capacity and Symbol Rate. In Proceedings of the Optical Fiber Communication Conference, San Diego, CA, USA, 11–15 March 2018.
10. Tan, M.; Iqbal, M.A.; Krzczanowicz, L.; Philips, I.D.; Harper, P.; Forysiak, W. Optimization of Raman Amplification Schemes for Single-Span High Data Rate Coherent Transmission Systems. In Proceedings of the Conference on Lasers and Electro-Optics, San Jose, CA, USA, 9–14 May 2021.
11. Al-Khateeb, M.; Tan, M.; Zhang, T.; Ellis, A. Combating Fiber Nonlinearity Using Dual-Order Raman Amplification and OPC. *IEEE Photonics Technol. Lett.* **2019**, *31*, 877–880. [CrossRef]
12. Krzczanowicz, L.; Phillips, I.D.; Iqbal, M.A.; Tan, M.; Harper, P.; Forysiak, W. Hybrid discrete Raman/EDFA design for broadband optical amplification in metro WDM systems. In Proceedings of the 19th International Conference on Transparent Optical Networks (ICTON), Girona, Spain, 2–6 July 2017; pp. 1–4.
13. Zhang, J.; Li, X.; Dong, Z. Digital Nonlinear Compensation Based on the Modified Logarithmic Step Size. *J. Lightwave Technol.* **2013**, *31*, 3546–3555. [CrossRef]
14. Nguyen, T.T.; Zhang, T.; Giacoumidis, E.; Ali, A.; Tan, M.; Harper, P.; Barry, L.; Ellis, A. Coupled Transceiver-Fiber Nonlinearity Compensation Based on Machine Learning for Probabilistic Shaping System. *J. Lightwave Technol.* **2021**, *39*, 388–399. [CrossRef]

Review

Distributed Raman Amplification for Fiber Nonlinearity Compensation in a Mid-Link Optical Phase Conjugation System

Mingming Tan [1,*], Paweł Rosa [2,*], Tu T. Nguyen [1,†], Mohammad A. Z. Al-Khateeb [1,†], Md. Asif Iqbal [1,‡], Tianhua Xu [3], Feng Wen [4], Juan D. Ania-Castañón [5] and Andrew D. Ellis [1]

[1] Aston Institute of Photonics Technologies, Aston University, Birmingham B4 7ET, UK; nguyentu.hcmuns@gmail.com (T.T.N.); elkhateeb.eng@gmail.com (M.A.Z.A.-K.); mdasif.iqbal@bt.com (M.A.I.); andrew.ellis@aston.ac.uk (A.D.E.)
[2] National Institute of Telecommunications, Szachowa 1, 04-894 Warsaw, Poland
[3] School of Engineering, University of Warwick, Coventry CV4 7AL, UK; tianhua.xu@warwick.ac.uk
[4] School of Information and Communication Engineering, University of Electronic Science and Technology of China, Chengdu 611731, China; fengwen@uestc.edu.cn
[5] Instituto de Óptica "Daza de Valdés", 28006 Madrid, Spain; jd.ania@csic.es
[*] Correspondence: m.tan1@aston.ac.uk (M.T.); p.rosa@il-pib.pl (P.R.)
[†] Current address: Infinera Corporation, Allentown, PA 18106, USA.
[‡] Current address: BT Applied Research, Adastral Park, Ipswich IP5 3RE, UK.

Abstract: In this paper, we review different designs of distributed Raman amplifiers which have been proposed to minimize the signal power profile asymmetry in mid-link optical phase conjugation systems. We demonstrate how the symmetrical signal power profiles along the fiber can be achieved using various distributed Raman amplification techniques in the single-span and more realistic multi-span circumstances. In addition, we show the theoretically predicted results of the Kerr nonlinear product reduction with different Raman techniques in mid-link optical phase conjugator systems, and then in-line/long-haul transmission performance using numerical simulations.

Keywords: Raman amplification; coherent fiber optic communications; optical phase conjugation

Citation: Tan, M.; Rosa, P.; Nguyen, T.T.; Al-Khateeb, M.A.Z.; Iqbal, M.A.; Xu, T.; Wen, F.; Ania-Castañón, J.D.; Ellis, A.D. Distributed Raman Amplification for Fiber Nonlinearity Compensation in a Mid-Link Optical Phase Conjugation System. *Sensors* **2022**, *22*, 758. https://doi.org/10.3390/s22030758

Academic Editors: Jiangbing Du, Yang Yue, Jian Zhao and Yan-ge Liu

Received: 17 December 2021
Accepted: 14 January 2022
Published: 19 January 2022

Publisher's Note: MDPI stays neutral with regard to jurisdictional claims in published maps and institutional affiliations.

1. Introduction

Mid-link optical phase conjugation (OPC) has been used to compensate both linear (e.g., chromatic dispersion) and the nonlinear (e.g., the Kerr nonlinearity) impairments of the optical fiber, which can significantly enhance the maximum transmission distance or the data capacity, particularly for a relatively long-haul transmission system [1–28]. There are a few limiting factors in the optical fiber link which constrain the efficiency of combating the nonlinear impairment in a mid-link OPC system, such as the chromatic dispersion slope and the signal power profile along the fiber [1–3]. Erbium-doped fiber amplifiers (EDFAs) are the most widely used amplification technique to compensate the loss in the optical fiber, but typically for a mid-link OPC system using the EDFAs in the link demonstrates that the either the maximum reach is not significantly extended, or the performance gain is modest [4–9], even when dispersion management is employed [10,11]. This is because of the lack of symmetrical signal power profiles before and after a mid-link OPC [12,13]. With increased availability of high power semiconductor pump lasers, and increased confidence in fiber power handling, Raman amplifiers are extensively used in unrepeatered submarine systems and terrestrial transmission systems [29,30], significantly improving the signal-to-noise ratio. Distributed Raman amplification (DRA) essentially uses the transmission fiber as the Raman gain medium and provides the signal amplification along the fiber, in comparison with an EDFA which is a lumped amplifier using a short Erbium-doped fiber as the gain medium. The design of DRA is highly flexible: Pump wavelength can be chosen and adjusted with fiber Bragg gratings (FBGs). A purposefully built DRA can also improve the power symmetry of the optical fiber link, enabling enhanced efficiency of nonlinearity compensation, and therefore the overall transmission performances [31–33]. Thus, the DRA not only provides an improved signal-to-noise ratio (SNR) without OPC, but also gives a large margin in the transmission performance (maximum reach, BER or data capacity) improvement when using an OPC to compensate a significant portion of nonlinear product in the transmission systems. In [12,13], K. Solis-Trapala et al. investigated the signal power symmetry and transmission performance of the bidirectional pumping over dispersion flattened non-zero dispersion-shifted fiber (NZDSF) in a mid-link OPC system, but the selected Raman scheme was restricted to first-order bidirectional pumping with similar pump power from both directions. The conclusion was that the shortest fiber length (25 km) gave the highest signal power profile symmetry. This optimized fiber length of 25 km is too short for realistic fiber spans in current optical transmission systems.

In this paper, we review various designs of distributed Raman amplification schemes aimed at improving the symmetry of the link for the transmission systems with mid-link optical phase conjugators. We show the optimized Raman amplification designs over single fiber span and multiple fiber spans (two) which can both demonstrate symmetry levels of above 93%. For the single-span link with a 50 km standard single mode fiber (SSMF) with backward Raman pumping only, the dual-order DRA can achieve 97% signal power profile symmetry and 39 dB nonlinear product compensation. This leads to more nonlinear product compensation, 12 dB higher, in comparison with conventional first-order DRA. For longer span links (e.g., 62 km), bidirectional Raman pumping is required to achieve good symmetry. A distributed Raman scheme based on the random distributed feedback laser architecture has been shown to maximize the signal power profile symmetry (97% symmetry) without introducing significant transmission performance penalty from relative intensity noise (RIN) of forward pumping. This proposed scheme can give 37.6 dB nonlinear product compensation, comparable to the 39 dB over 50 km dual-order backward pumping Raman scheme. Furthermore, for multi-fiber-span (i.e., 2 × 50 km SSMF) links, the impact of loss between the spans is crucial when optimizing the symmetry. The conventional dual-order backward-pumped DRA can give only ~66% signal power profile symmetry, achieving ~17 dB nonlinear product compensation. An over-pumped first span using the net gain to compensate the loss between spans would help improve the symmetry to 81% leading to a 25 dB nonlinear product reduction. The best solution is that a 25 cm erbium-doped fiber (EDF) is embedded in the conventional dual-order DRA, but the Raman pumps are used to pump the EDF and so compensate the loss between the two spans, which enables the overall link symmetry to be optimized to 93.9% providing up to 32 dB nonlinear product compensation with 50 km per span. In addition, we demonstrate that in a relatively short transmission system (100–200 km), the higher signal power profile symmetry, the higher the nonlinear threshold (up to 9 dB) when using OPC. We also show, in the numerical simulations, a Q^2 factor improvement of approximately 8 dB in the long-haul transmission systems (2000 km) using the optimized EDF-assisted Raman amplified spans in the link.

2. Optimized Distributed Raman Amplification Design over Single Fiber Span

As the stimulated Raman scattering allows a Stokes shift (13 THz) from the pump to the signal which gives more flexibility for the choice of the pump wavelengths and the Raman gain fiber, Raman amplification is highly configurable. For example, in Reference [30], a cascaded third-order Raman pump was used in an unrepeatered transmission experiment. In Reference [34], the authors presented a sixth-order Raman pump configuration. However, the commonly used Raman schemes are based on first-order and second-order Raman amplifiers which have relatively configurations, higher pump-signal conversion efficiency and consequently lower cost [35–43]. Using only forward Raman pumping is generally not feasible for transmission systems because firstly, it introduces significant RIN-related penalty to the transmitted signal from the pump, and secondly the signal suffers high fiber nonlinearity due to high signal power near the input sections of the transmission span [35–43]. Therefore, in this paper, we mainly focus on the designs of first-order and second-order distributed Raman amplification schemes using backward or bidirectional pumping which requires the RIN penalty mitigation technique.

2.1. Distributed Raman Amplification with Backward Pumping Only

Figure 1a shows the schematic diagram of the conventional first-order backward-pumped DRA, including the 50.4 km SSMF pumped by a fully depolarized fiber laser at 1455 nm. Because the fiber length is only ~50.4 km mainly for short reach or multiple spans in long-haul transmission systems, the span loss was fully compensated which means that the pump power was commonly set to achieve zero dB net gain [43–48]. Figure 1b shows both experimentally measured and numerically simulated signal power profiles along the fiber using the modified optical time domain reflectometer (OTDR) technique [44]. The symmetry/asymmetry level of the signal power profile is based on the calculation method in [12,13], which gave 89% symmetry (11% asymmetry) for this Raman scheme.

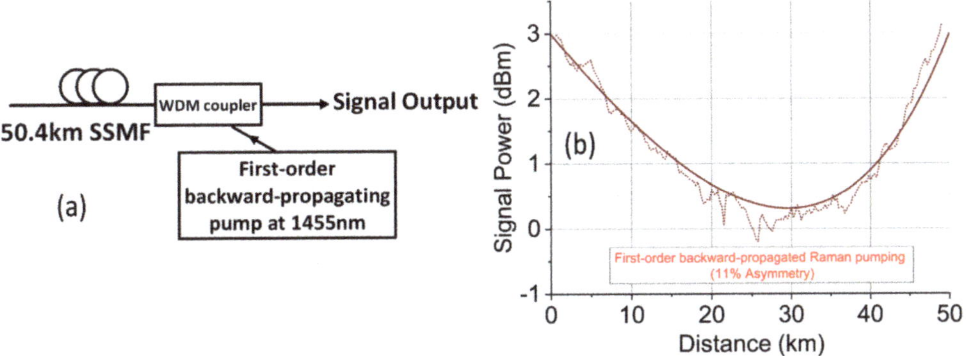

Figure 1. (**a**) First-order backward pumping; (**b**) signal power profiles along the fiber.

Figure 2a shows a dual-order backward-pumping DRA scheme including both a 1365 nm Raman fiber laser and a 1455 nm pump seed [49,50]. Figure 2b shows the simulated and experimentally measured signal power profiles along the fiber. The use of the dual-order BW-pumping scheme can improve the symmetry of signal power profile, but it requires the optimization of the first- and second-order pump power [23]. As shown in Figure 2b, to achieve 97% symmetry, the first- and second-order pump power was set to ~100 and ~330 mW respectively. When the second-order pump power is increased beyond this optimum, the signal power symmetry degrades to 92% with ~33 and ~600 mW first- and second-order pump powers, respectively. This asymmetry accumulation was mainly because the signal power was increased near the signal output, as the signal gain was pushed into the middle of the fiber span using higher second-order pump power [23], whilst for lower second-order pump powers, the asymmetry approaches that of the first-order pumped configuration.

As the alternative to the dual-order Raman scheme, the pump seed at 1455 nm can be replaced by a highly reflective fiber Bragg grating (FBG) as shown in Figure 3a [35]. In this case, a random distributed feedback (DFB) fiber laser at 1455 nm was generated due to the distributed Rayleigh backscattering (originating from the SSMF) and the fixed reflection of the FBG [35–37]. This scheme is more cost-effective in comparison with the dual-order scheme, as no active pump at 1455 nm is required. However, in terms of flexibility when optimizing the signal power symmetry, a random DFB fiber laser is not flexible compared to independent separate pumps as a minimum second-order pump power is determined by the grating reflectivity and the stimulated Brillouin scattering (SBS) coefficient. The same parameters also determine the ratio of first to second-order pump power. For a typical grating reflectivity of 99%, the power symmetry was simulated to be only 75% for the 50.4 km SSMF span.

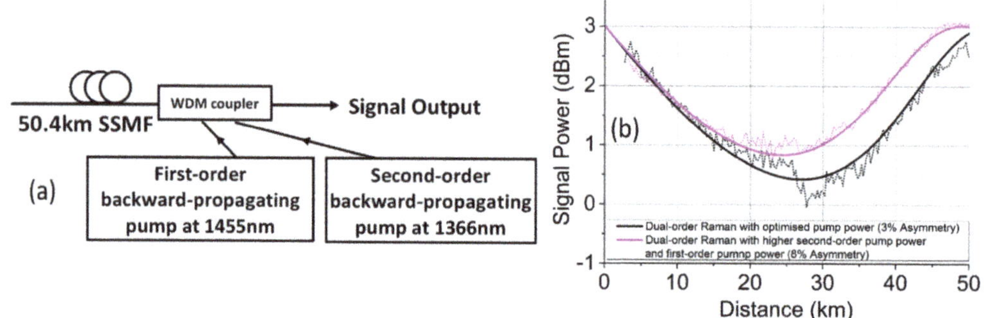

Figure 2. (**a**) Dual-order backward pumping; (**b**) signal power profiles along the fiber.

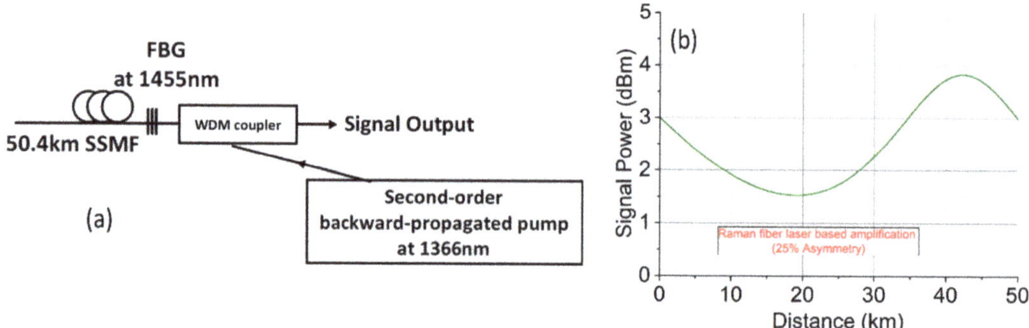

Figure 3. (**a**) Raman-fiber-laser-based amplification with second-order pumping; (**b**) simulated signal power profiles.

We have theoretically predicted the power of the nonlinear Kerr product generated by two co-polarized CW lasers (3 dBm each) along a 50.4 km backward-pumped DRA system with and without mid-link OPC. Figure 4 shows the configuration used to predict the power of the Kerr nonlinear product as a function of frequency separation of the two lasers [23]. Figure 5 demonstrates the nonlinear product power versus laser frequency separation using the three different Raman schemes described above without and with mid-link OPC. For the conventional first-order DRA, the nonlinear product power without OPC was up to −11.5 dBm at low frequency ranges. With OPC, the peak Kerr product power was suppressed by 27 dB. Using an optimized dual-order Raman scheme would improve the symmetry of signal power distributions and therefore further decrease the nonlinear Kerr product using the mid-link OPC. In Figure 5b, using the non-optimized pump power configuration (600 mW second-order pump power), the peak Kerr product power reduction was improved to ~30 dB, but once the pump power was optimized to maximize the symmetry (330 mW second-order pump power), the Kerr product reduction was increased to 39 dB. The random-fiber-laser-based scheme gave the lowest symmetry, and therefore the reduction in Kerr product power with OPC was limited to ~20 dB, as illustrated in Figure 5c.

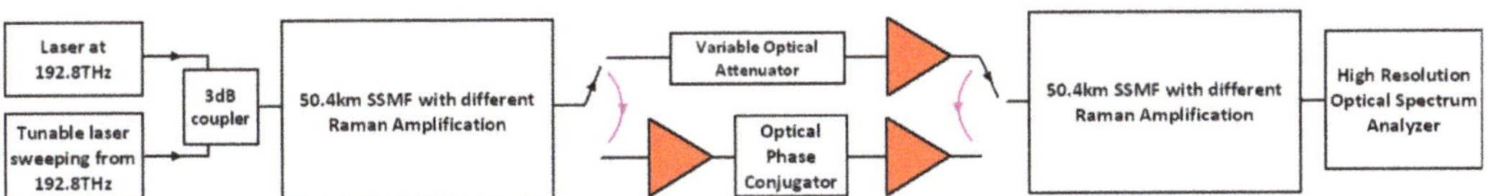

Figure 4. Schematic diagram of nonlinear product measurement using dual-order DRA in mid-link OPC systems.

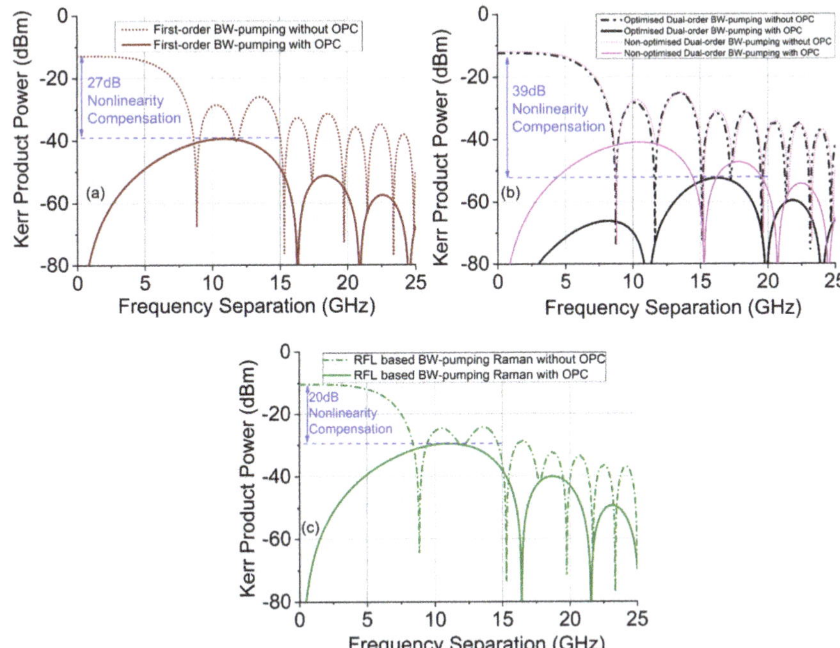

Figure 5. Theoretical predication of nonlinear Kerr product power as a function of frequency separation between the two CW lasers over 50 km SSMF. (**a**) First-order Raman amplification. (**b**) Dual-order Raman amplification with optimized pump power and non-optimized pump power). (**c**) Random Raman-fiber-laser-based amplification.

An inline coherent transmission experiment was conducted by replacing the two CW lasers shown in Figure 4 with a 256 Gb/s DP-16QAM (32 GBaud) signal (centered at 194.8 THz). After the transmission, the signal or its conjugate (centered at 194.65 THz) was amplified with an EDFA as the receiver amplifier before being detected by a polarization-diverse coherent receiver (100 GSa/s, analog bandwidth of 33 GHz). Commercial digital signal processing (DSP) software was used to process the captured data from the scope with Q^2 factors calculated from the bit-error-rate of 500,000 bits.

As shown in Figure 6, ~5 dB improvement in the nonlinear threshold and ~7 dB increases in launch power for a fixed Q^2 in the nonlinear region are achieved when using the nearly perfect (~97%) signal power symmetry provided by dual-order BW-pumping DRA. The optimum Q^2 factor was reduced because of the short transmission distance (100 km) and the additional ASE noise added by the inclusion of the OPC. However, using the optimized OPC in a long-haul transmission system has been proven to significantly improve the system performance, since the transceiver and OPC noise are negligible compared with the accumulated link noise which is the majority of the linear noise [23].

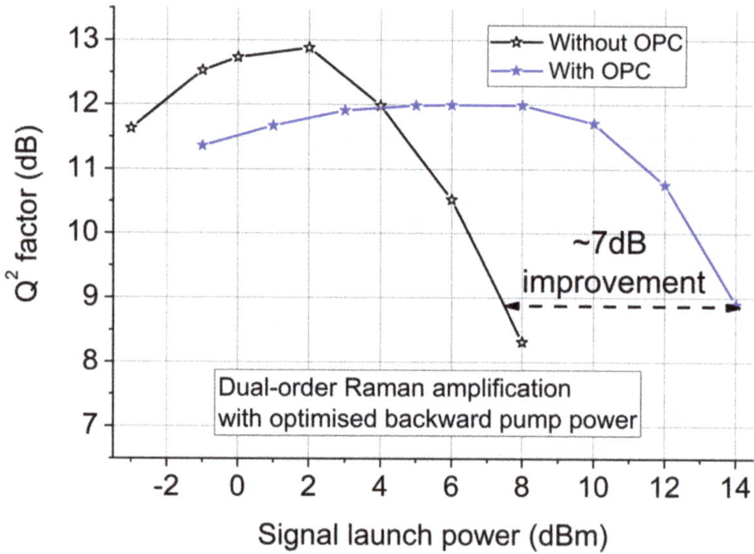

Figure 6. Experimentally measured Q^2 factors versus signal launch power in inline transmission systems without/with OPC using the optimized dual-order DRA scheme over 2×50 km SSMF.

2.2. Distributed Raman Amplification with Bidirectional Pumping

In our optimization of signal power profiles for mid-link OPC we consider three different distributed Raman amplification schemes, all of them bidirectionally pumped. Simulations are performed for each configuration, obtaining signal power excursion for different pump power ratios and span lengths using the tried and tested model fully described in [38] with the boundary conditions corresponding to each of the cases under consideration and assuming fully depolarized pumps as well as room temperature operation. Noise calculations are based on a 0.1 nm bandwidth. The coefficients for Raman gain and attenuation at the different wavelengths involved were extracted from measurements for SSMF [38], whereas the Rayleigh backscattering coefficients at 1366 and 1455 nm and the frequency of the signal are assumed to be 1.0×10^{-4}, 6.5×10^{-5} and 4.5×10^{-5} km^{-1}, respectively.

The first case considered (Figure 7a) corresponds to a conventional bidirectionally pumped first-order Raman amplifier, with pumps at 1455 nm that amplify the signal through the first Stokes shift.

The second case corresponds to an ultra-long Raman fiber laser (URFL) amplifier (Figure 7b), that provides second-order pumping from single-wavelength pumps. In such an amplifier, the initial Raman fiber laser pumps operate at 1366 nm, that is, downshifted by two Stokes shift with respect to the frequency of the signal. Highly reflective (99%) fiber Bragg gratings (FBGs) centered at 1455 nm with a bandwidth of 200 GHz are located at both ends of the transmission line to back-reflect the first Stokes-shifted radiation at 1455 nm into the long cavity. Once a threshold of ~0.8 W pump power is reached, the cavity forms a stable ultra-long laser that at that amplifies the signal around 1550 nm. This approach presents the advantage of having modifiable gain bandwidth and profiles by selecting appropriate FBGs, instead of requiring an active seed at the intermediate Stokes. In this case the reflectivity of the FBGs was chosen to be high to maximize pump-to-signal power conversion efficiency [38–41].

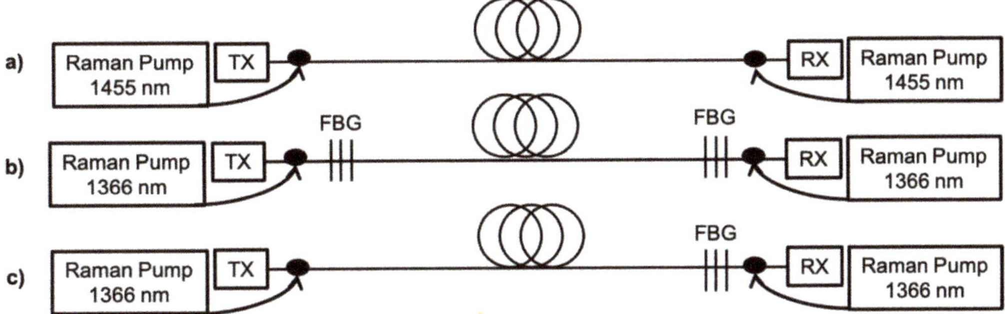

Figure 7. Schematic design of (**a**) first-order Raman, (**b**) second-order ultra-long fiber laser and (**c**) second-order random DFB Raman laser amplifiers.

The third and final approach uses a random distributed feedback Raman laser amplifier (Figure 7c). This is another fiber laser amplifier similar in part to the scheme shown in Figure 3. This is significantly different from a closed cavity with two FBGs. The scheme is essentially bidirectional Raman pumping using a half-open-cavity design with a single high-reflectivity FBG at 1455 nm located at the end of the span that reflects the 1455 nm Stokes in the backward direction. The second-order pump in front of the span, does not create a seed traveling in forward direction (either by inserting an FBG or an active seed) but rather amplifies the seed created at the end of the span. The lack of an FBG on the side of the forward pump reduces the RIN transfer to the 1455 nm Stokes in exchange for a reduction in conversion efficiency in comparison to the other two proposed setups [35]. This reduced interaction between the signal and the forward pumping is particularly important, as the RIN transfer from high-power forward pumps can be a limiting factor in data transmission [42,43].

In order to perform a comparison of signal power asymmetry between the three proposed configurations, we simulated power profiles of a single channel at 1545 nm with a fixed launch power of 0 dBm. For each value of the forward pump power (FPP) and span lengths ranging from 40 to 100 km, the backward pump was adjusted to provide 0 dB net gain. Signal power asymmetry was calculated as in [12,13]. Full results using random DFB laser amplification (Figure 7c) are shown in Figure 8. Data provide a broader picture of the asymmetry evolution in transmission over a broad range of forward pump powers up to 2.5 W and lengths with the optimal backward pumping (0 dB net gain). The lowest asymmetry point is found to be at 62 km, with signal power asymmetry just below 3% (97% symmetry). Further optimization is possible based on simultaneous ASE noise minimization and nonlinearity compensation [29].

In order to confirm the simulation results, in Figure 9 we compare the simulated prediction and experimentally measured asymmetry vs. forward pump power split for a signal at 1545 nm in a 60 km span (the particular length was chosen due to availability of a SSMF fiber reels). The discrepancies between measurement and simulation are attributable mainly to the noisy experimental power profiles, as well as the mismatch of Raman gain and attenuation coefficients, since for consistency with previous simulations we used standard values for SSMF instead of measured coefficients.

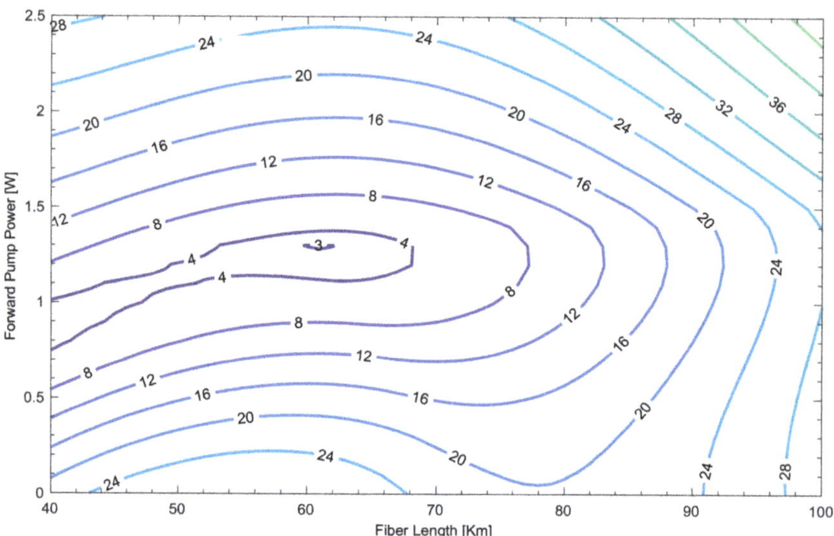

Figure 8. Signal power asymmetry (%) as a function of different span lengths and pump powers (backward pump power was adjusted to give 0 dB net gain).

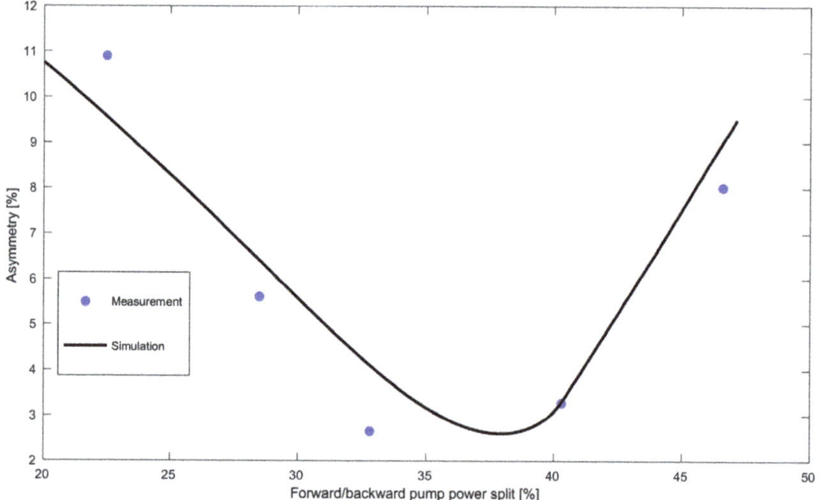

Figure 9. Asymmetry dependence on the forward pump power split measured at the central wavelength at 1545 nm in a 60 km span.

Finally we numerically compare three amplification schemes shown in Figure 7 using the same method as described in Figure 8: In each case we simulate all possible FPPs adjusting backward pump to give 0 dB net gain and choose the point with the best asymmetry level (with the favor of a lower forward pump power in case if the same asymmetry is achieved for two different FPP). The results are summarized in Figure 10 below. The random DFB Raman laser amplification setup (red) achieved the lowest asymmetry levels for span lengths above 58 km. The URFL amplification option displayed better symmetry for lengths between 40 and 58 km, with optimal forward/backward pumping power ratios close to 1 in spans of up to 50 km but requiring higher contribution from the backward pump as span length grows. The random DFB configuration requires a higher contribution of the backward pump for lengths of up to 30 km, but forward/backward pump power ratio achieves close to 1 for longer spans. Optimal symmetry in first-order Raman amplification is found for backward pumping only.

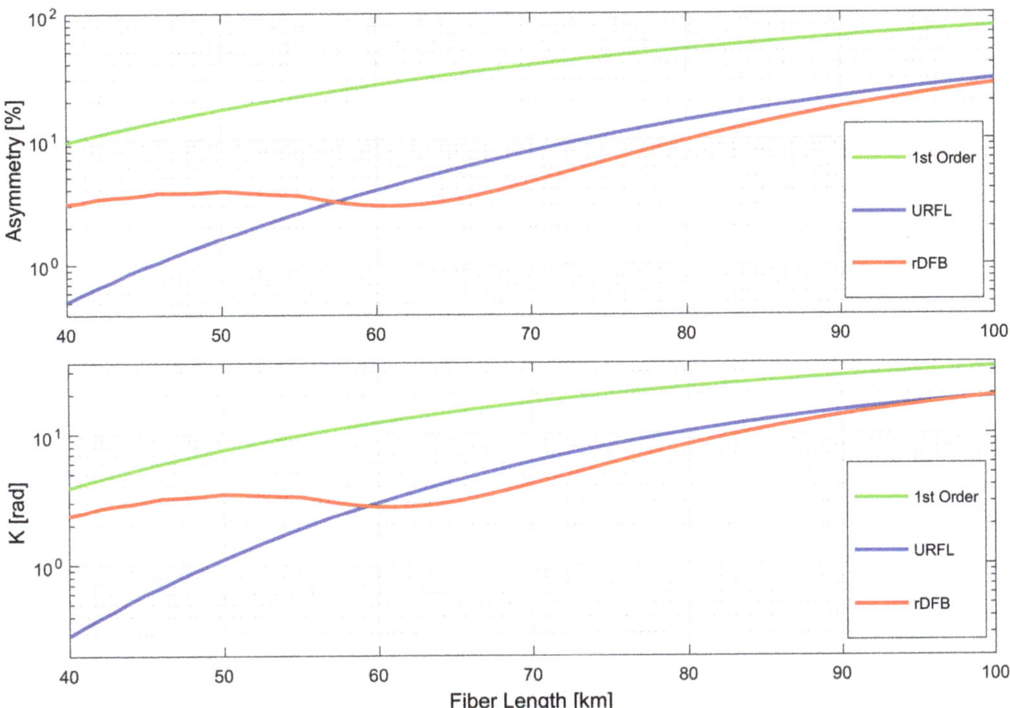

Figure 10. Lowest signal power asymmetry for a given length with pump powers adjusted to give zero net gain (**top**) and the accumulated residual phase shift (**bottom**) for a given amplification scheme.

The bottom of Figure 10 shows the potential accumulated residual phase shift after optimal OPC, defined as the product of the optimal asymmetry at a given distance and the corresponding nonlinear phase shift.

The combined results for high symmetry and residual phase shift results, together with resiliency to forward- pumping RIN in coherent transmission applications [29], suggest that a bidirectionally pumped random DFB laser with a single grating seems to be the best option, performance-wise, for amplification in long spans with OPC. Thus, we chose this as the option for further study.

Figure 11 shows the Kerr product reduction versus frequency separation using almost symmetrical signal power profile (bidirectional pumping with random distributed feedback fiber laser). In Figure 11, without OPC, the nonlinear product power was up to −6.8 dBm near the low frequency range, but with OPC, the nonlinear product power was decreased to −44.4 dBm at 18 GHz. There was 37.6 dB nonlinear product degradation with the 97% symmetrical signal power profiles over 62 km SSMF, which was comparable with the optimized dual-order backward pumping Raman amplification over 50 km SSMF.

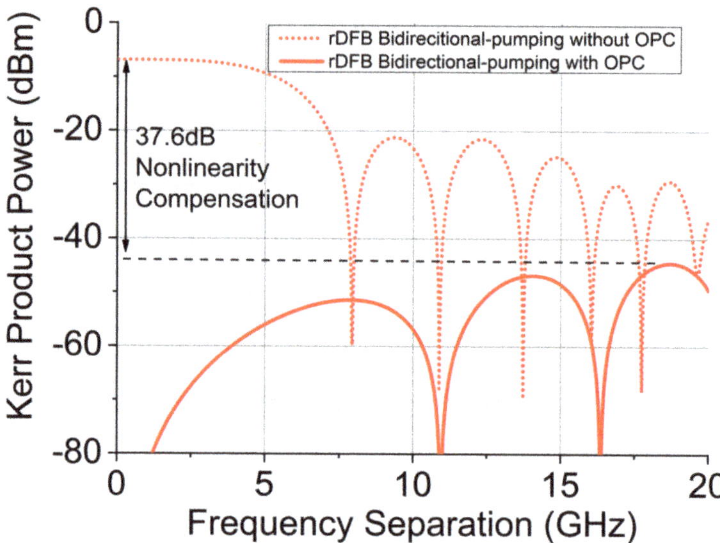

Figure 11. Theoretical predication of nonlinear Kerr product power as a function of frequency separation between the two CW lasers using rDFB bidirectional-pumping Raman amplification over 62 km SSMF span.

3. Optimized Distributed Raman Amplification Design over Multiple Fiber Spans

We have studied the optimization of Raman amplification schemes with backward pumping only over multiple fiber spans in [2], and this has been the only study so far regarding multiple fiber span links.

In order to investigate the impact of multi-span DRA schemes on the nonlinearity compensation, we demonstrated the signal profiles of three dual-order DRA schemes over 2×50.4 km SSMF spans and illustrated four-wave mixing (FWM) conversion efficiencies and inline/long-haul transmission performances.

Scheme 1 was the dual-order DRA with backward pumping which was the same as Figure 2 in each individual span that included both the first-order pump seed at 1455 nm and the second-order pump at 1366 nm [49,50]. The gain of the dual-order DRA compensated only the loss from the SSMF, and the losses of the passive components (e.g., signal pump combiners, isolators) was not compensated. In scheme 2, the experimental setup was the same as scheme 1, and the pump power for the second span also remained the same, but the 1.5 dB loss due to fiber attenuation and passive components between the two spans was compensated by more Raman power from the first span. This means that the signal output power from the first SSMF was overpowered by ~1.5 dB. In scheme 3, to compensate the loss from passive components without sacrificing the symmetry of the two-span link, a 25 cm EDF (Fibrecore M-12(980/125)) was used between the SSMF and the WDM coupler at the end of the first span as illustrated in Figure 12b, which enabled 1.5 dB amplification within this 25 cm erbium-doped fiber.

Figure 13 shows the experimentally measured and numerically simulated signal power profiles along the fiber for all three DRA schemes using the modified OTDR setup [44]. In scheme 1, the pump power was 330 mW at 1366 nm and 100 mW at 1455 nm. In the two-span link, ~1.5 dB signal power difference between two spans existed which reduced the overall signal power profile symmetry across the two spans to 65.8%, a drop dominated by the launch power difference.

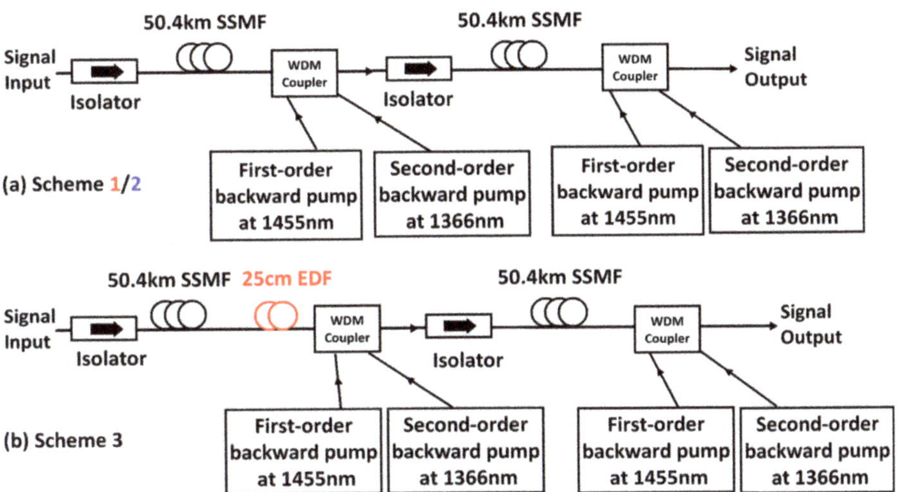

Figure 12. (**a**) Schemes 1 and 2: Dual-order BW-propagated pumping DRA over 2 × 50.4 km SSMF (including two pump power settings). (**b**) Scheme 3: EDF-assisted dual-order BW-propagated pumping DRA over 2 × 50.4 km SSMF.

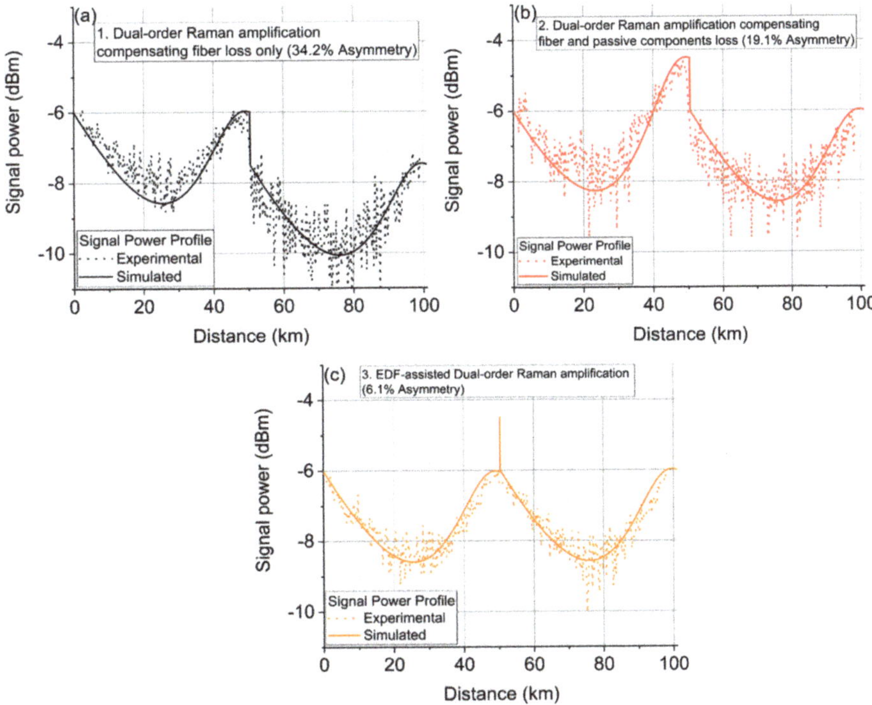

Figure 13. Experimentally measured and simulated signal power profiles along the fiber with three Raman amplification schemes; (**a**) DRA compensating only the fiber loss. (**b**) DRA compensating the fiber and the passive components loss between the two spans. (**c**) EDF-assisted DRA compensating fiber and the passive components loss between the two spans.

In scheme 2, the pump power in the first span was 330 mW at 1366 nm and 115 mW at 1455 nm, higher than in scheme 1 to compensate the 1.5 dB loss between the spans. Matching launch powers thus improved the symmetry to 80.9%.

In scheme 3, the pump power before passing through the EDF in the first span was 330 mW at 1366 nm and 108 mW at 1455 nm. The 1.5 dB EDF amplification was powered by both Raman pumps (primarily by 1455 nm; however, the power from 1366 nm pump would be transferred to 1455 nm and contribute to the gain at low C band simultaneously) and therefore compensated the loss of passive components between spans [35]. This is different from the conventional hybrid Raman/EDFA: Instead of using 980/1480 nm laser pumps to invert the EDF fiber [51] we relied on the Raman pumps at 1366 and 1455 nm. By matching both launch power and signal evolution in the two-span Scheme enables the signal power symmetry level of 93.4% over 2 × 50.4 km spans. The symmetry level of 93.4% over two spans is also comparable to the best symmetry level of 97% over a single 50 km span using an optimized dual-order Raman amplification shown in [2].

The theoretically predicted nonlinear product power is shown in Figure 14 as a function of frequency separation of the two lasers using the three DRA schemes. With no OPC, the nonlinear Kerr product power for all the three DRA schemes was up to −16 dBm in the low frequency region. However, because of the poor link symmetry caused by the signal power degradation between spans, scheme 1 showed the least Kerr power reduction of only ~17 dB with a mid-link OPC. Higher Kerr product power compensation (~25 dB) was achieved using the first over-pumped span. The EDF-assisted scheme generated ~1.5 dB gain to compensate the loss between spans and therefore achieved excellent signal power symmetry simultaneously for both single-span and two-span links, which showed the highest compensation (up to 32 dB) in nonlinear Kerr product power.

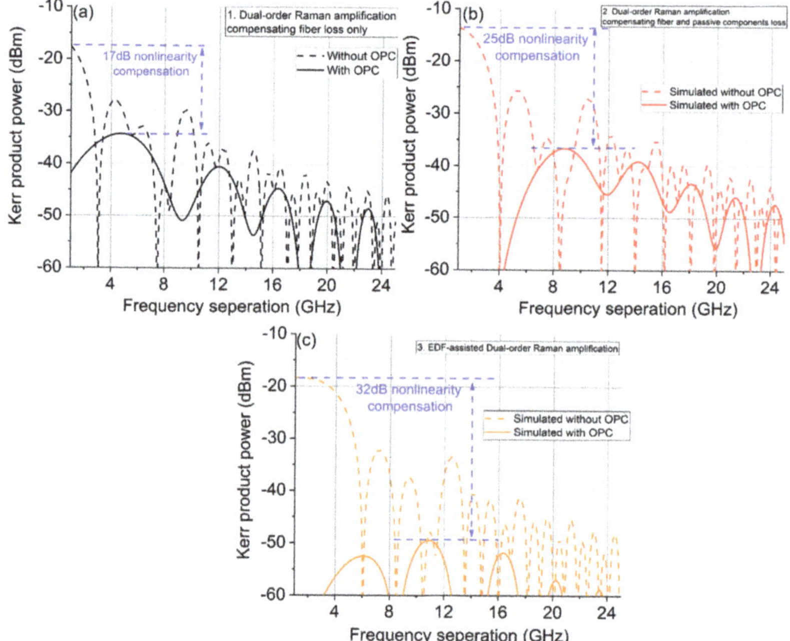

Figure 14. Theoretically predicated nonlinear product frequency separation without/with a mid-link OPC using different DRA schemes over 2 × 50 km spans. Dotted lines—no OPC, solid-with OPC. (**a**) No account of loss, (**b**) accounting for loss with excess Raman gain, (**c**) accounting for loss with EDF.

A numerical simulation was performed based on 2 × 2 spans (50.4 km per span, which makes ~200 km in total) using a 200 Gb/s DP-16QAM (32 GBaud, 256 Gbit/s line rate, 2^{16} PRBS length, 0.1 roll-off factor) signal centered at 194.8 THz with the three different DRA schemes (DP-16QAM signal). The nonlinear Schrödinger equations (Manakov equations) were solved using the well-known split-step Fourier method [52–54] with a step size of 0.1 km, in which the signal power profiles shown in Figure 13 were used. The noise from each amplifier was modeled as Gaussian noise and added to the signal after each step (0.1 km) [2] to ensure that parametric noise amplification was correctly captured. The ASE noise of EDFAs at the transmitter, the OPC and the receiver were considered in the simulations (−140 dBm/Hz noise power density). More details about the simulation parameters can be found in [2].

In Figure 15, the EDF-assisted scheme 3 shows a maximum launch power improvement of 9 dB at a fixed Q^2 factor in the nonlinear regime, exceeding the improvement observed for the conventional DRA schemes 1 and 2 (by 4 and 2 dB), respectively. This is due to the nearly perfect signal power symmetry (>93% symmetry) from the EDF-generated gain compensating the loss between spans. However, as the overall transmission distance was ~200 km, the noise from the Raman-amplified link was limited, and therefore the results were dominated by optical noise from EDFAs in the transmitter, OPC and receiver, which obscured the Q^2 factor benefit introduced by the nonlinearity compensation from the mid-link OPC and contributed to the small reduction in optimum Q^2 factor with mid-link OPC [2].

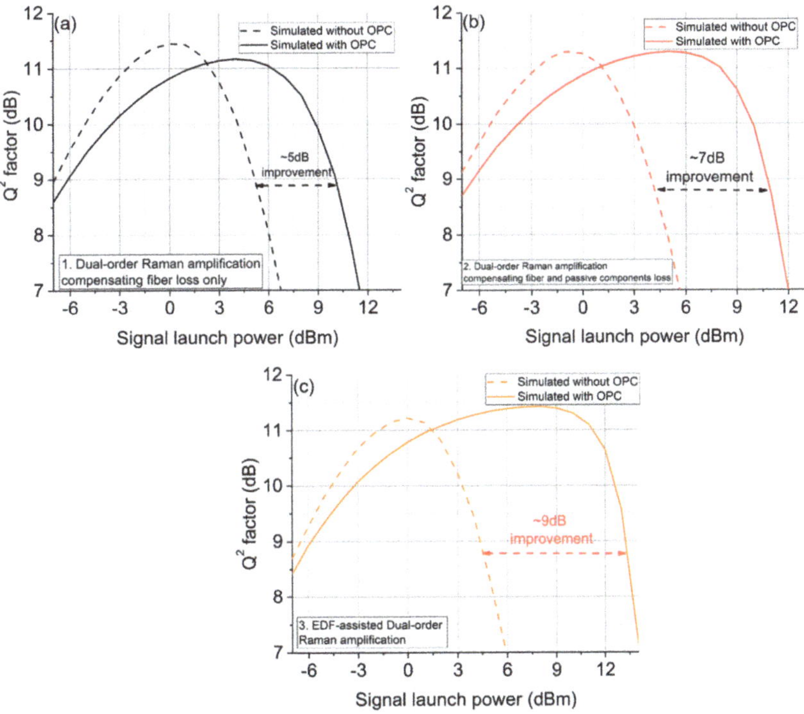

Figure 15. Q^2 factors versus signal launch power in inline transmission systems without/with OPC using the optimized dual-order DRA scheme at 200 km. Dotted lines—no OPC, solid—with OPC. (**a**) No account of loss, (**b**) accounting for loss with excess Raman gain, (**c**) accounting for loss with EDF.

Figure 16 shows simulated transmission performances at 2000 km using the EDF-assisted Raman Scheme. As expected, the most symmetrical Raman scheme (Figures 13c and 15c) gives ~8dB Q^2 factor improvement (Q^2 factor is defined as 20log10(sqrt(2).*erfcinv(2.*(BER)). In addition, the EDFA noise at the transmitter and receivers was deliberately removed for the EDF-assisted Raman scheme to evaluate the impact of such noise. In this case over 2000 km, the impact of the noise from EDFAs at the transmitter and receiver becomes negligible as the Q^2 factor differences are very small. Thus, the accumulated noise from the Raman amplified link is more dominant compared with transceiver noise (fundamentally limited by higher-order parametric noise), and therefore the benefit of the Q^2 factor improvement can be revealed for long-haul transmission systems with an optimized mid-link OPC [21,55]. Therefore, using the symmetrical EDF-assisted Raman link in long-haul transmission systems can improve the fiber nonlinearity compensation efficiency and the transmission performance in a mid-link OPC system.

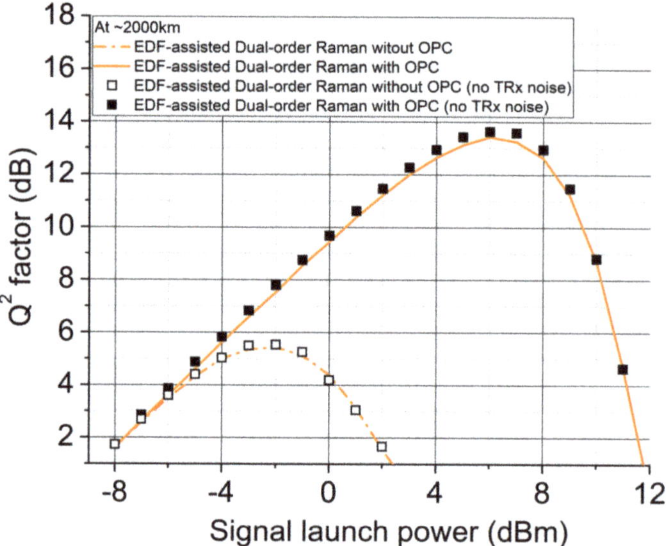

Figure 16. Q^2 factors versus signal launch power in inline transmission systems without/with OPC using the optimized dual-order DRA scheme.

4. Discussion

From our analysis in Sections 2 and 3, to achieve the best symmetry and higher efficiency of combating fiber nonlinearity, different Raman schemes have to be considered for different span lengths. A table (Table 1) summarizing the span symmetry with Raman pumping schemes at different span lengths is demonstrated below. It is shown that for the short span length of 25 km, the first-order bidirectional Raman pumping was sufficient to achieve 97% signal power symmetry, but the length of 25 km was very short for OPC-based application (e.g., long-haul transmission systems). In addition, for this short length, the symmetry changes will be relatively small when using different Raman schemes, and bidirectional Raman pumping will introduce significant RIN-replated penalty. For the span length of around 50 km, optimized dual-order Raman pumping is required to achieve 97% span symmetry. However, for longer span length, bidirectional second-order Raman pumping would be needed as the signal gain can be generated from the start of the span. Thus, the scheme based on a random fiber laser with bidirectional second-order pumping without introducing RIN penalty was demonstrated to achieve 97% span symmetry at 62 km. Further extending the span length to 100 km, higher order bidirectional pumping would be required, but given the RIN penalty introduced from conventional Raman pumps, we had to stick to RIN-penalty-free bidirectional pumping based on a random fiber laser. In this case, the optimum span symmetry dropped to 72% at 100 km. Alternatively, we could break the 100 km into two 50 km spans, and then the problem became how to leverage the loss between the two spans. We used the EDF with Raman pumps to account for the loss between spans and improved the span symmetry from 72% (single span) to 93% (two spans).

Table 1. Summary of the best span symmetry with corresponding Raman schemes at different span lengths.

Span Lengths	Raman Pumping Schemes	Optimum Signal Power Profile Symmetry
25 km	First-order bidirectional pumping	97% [12]
50 km	Dual-order backward pumping	97% [23]
62 km	Second-order bidirectional pumping with random fiber laser	97% [29]
100 km	Second-order bidirectional pumping with random fiber laser	72% [29]
2 × 50 km	Dual-order backward pumping with EDF in the first span, no EDF in the second span	93% [2]

5. Conclusions

We review the application of distributed Raman amplifiers with different designs for nonlinearity compensation in mid-link OPC systems. We demonstrate for single-span system with mid-link OPC that a dual-order backward-pumped Raman scheme can efficiently compensate the nonlinearity given that the pump powers are optimized to maximize the signal power profile symmetry. We show that using optimized pump powers can achieve up to 97% symmetry and 39 dB nonlinear product power reduction using a mid-link OPC. For longer span length, bidirectional Raman pumping is required to maintain a similar level of symmetry. We demonstrate that a random fiber laser amplifier is the most suitable solution for mid-link OPC WDM systems using span lengths between 60 and 100 km with the best performance at the distance of 62 km, demonstrating the Kerr product reduction up to 37.6 dB.

For multiple span systems, the optimized configurations (utilizing a 25 cm EDF) improve signal power profile symmetry and consequently enhance fiber nonlinearity compensation efficiency. This technique can compensate the loss of passive components between the spans and therefore maximize the overall signal power symmetry up to 93% in realistic multi-fiber-span link in a cost-effective manner. We demonstrate that, in the multi-span link with a mid-link OPC, using this scheme shows ~32 dB nonlinear product compensation that is at least 7 dB higher than conventional dual-order Raman schemes. We also show that, for nearly symmetrical signal power profiles, the Raman schemes in both the single-span and two-span systems give a 9 dB enhancement of the nonlinear threshold in the 200 Gb/s DP-16QAM transmission system using a mid-link OPC.

Author Contributions: M.T. and P.R. proposed the concept and initiated the study. M.T., P.R., T.T.N., M.A.Z.A.-K., M.A.I. and J.D.A.-C. carried out numerical simulations. M.T., M.A.Z.A.-K. and F.W. performed the experiments. M.T., P.R., T.T.N., M.A.Z.A.-K., F.W. and T.X. conducted analytical calculations. J.D.A.-C. and A.D.E. supervised the studies. The paper was written by M.T. and P.R. All authors have read and agreed to the published version of the manuscript.

Funding: This research was funded by the UK Engineering and Physical Sciences Research Council (EPSRC) Grant EP/S003436/1 (PHOS), EP/V000969/1 (ARGON), EP/S016171/1(EEMC), the Polish Ministry of Science and Higher Education Grant 12300051, EU Horizon 2020 Project 101008280, RTI2018-097957-B-C33 (ECOSYSTEM) funded by Spanish MCIN/AEI/10.13039/501100011033, Research and innovation Programme SINFOTON2-CM (S2018/NMT-4326) co-financed by ESF funds, and the National Natural Science Foundation of China Grant 61975027.

Institutional Review Board Statement: Not applicable.

Informed Consent Statement: Not applicable.

Data Availability Statement: Original data are available at Aston Research Explorer (https://doi.org/10.17036/researchdata.aston.ac.uk.00000534 (accessed on 16 December 2021)).

Acknowledgments: The authors would like to thank Paul Harper and Lukasz Krzczanowicz for insightful discussions and Sterlite Technologies and Finisar for industrial support.

Conflicts of Interest: The authors declare no conflict of interest.

References

1. Minzioni, P.; Cristiani, I.; Degiorgio, V.; Marazzi, L.; Martinelli, M.; Langrock, C.; Fejer, M.M. Experimental Demonstration of Nonlinearity and Dispersion Compensation in an Embedded Link by Optical Phase Conjugation. *IEEE Photonics Technol. Lett.* **2006**, *18*, 995–997. [CrossRef]
2. Tan, M.; Nguyen, T.T.; Rosa, P.; Al-Khateeb, M.A.Z.; Zhang, T.T.; Ellis, A.D. Enhancing the Signal Power Symmetry for Optical Phase Conjugation Using Erbium-Doped-Fiber-Assisted Raman Amplification. *IEEE Access* **2020**, *8*, 222766–222773. [CrossRef]
3. Rosa, P.; Rizzelli, G.; Ania-Castañón, J.D. Link optimization for DWDM transmission with an optical phase conjugation. *Opt. Express* **2016**, *24*, 16450–16455. [CrossRef] [PubMed]
4. Du, L.B.; Morshed, M.M.; Lowery, A.J. Fiber nonlinearity compensation for OFDM super-channels using optical phase conjugation. *Opt. Express* **2012**, *20*, 19921. [CrossRef] [PubMed]
5. Huang, C.; Shu, C. Raman-enhanced optical phase conjugator in WDM transmission systems. *Opt. Express* **2018**, *26*, 10274. [CrossRef]
6. Kaminski, P.; Da Ros, F.; Clausen, A.T.; Forchhammer, S.; Oxenløwe, L.K.; Galili, M. Improved nonlinearity compensation of OPC-aided EDFA-amplified transmission by enhanced dispersion mapping. In Proceedings of the 2020 Conference on Lasers and Electro-Optics (CLEO), San Jose, CA, USA, 10–15 May 2020.
7. Yoshima, S.; Sun, Y.; Liu, Z.; Bottrill, K.R.H.; Parmigiani, F.; Richardson, D.J.; Petropoulos, P. Mitigation of Nonlinear Effects on WDM QAM Signals Enabled by Optical Phase Conjugation with Efficient Bandwidth Utilization. *J. Lightwave Technol.* **2017**, *35*, 971–978. [CrossRef]
8. Sackey, I.; Schmidt-Langhorst, C.; Elschner, R.; Kato, T.; Tanimura, T.; Watanabe, S.; Hoshida, T.; Schubert, C. Waveband-Shift-Free Optical Phase Conjugator for Spectrally Efficient Fiber Nonlinearity Mitigation. *J. Lightwave Technol.* **2018**, *36*, 1309–1317. [CrossRef]
9. Sackey, I.; Da Ros, F.; Fischer, J.K.; Richter, T.; Jazayerifar, M.; Peucheret, C.; Petermann, K.; Schubert, C. Kerr Nonlinearity Mitigation: Mid-link Spectral Inversion versus Digital Backpropagation in 5x28GBd PDM 16-QAM Signal Transmission. *Opt. Express* **2014**, *22*, 27381–27391. [CrossRef]
10. Jasen, S.L.; van den Borne, D.; Spinnler, B.; Calabro, S.; Suche, H.; Krummrich, P.M.; Sohler, W.; Khoe, G.-D. Optical phase conjugation for ultra long-haul phase-shift-keyed transmission. *J. Lightwave Technol.* **2006**, *24*, 54–64. [CrossRef]
11. Pelusi, M.D. WDM signal All-optical Precompensation of Kerr Nonlinearity in Dispersion-Managed Fibers. *IEEE Photonics Technol. Lett.* **2013**, *25*, 71–73. [CrossRef]
12. Solis-Trapala, K.; Pelusi, M.; Tan, H.N.; Inoue, T.; Namiki, S. Optimized WDM Transmission Impairment Mitigation by Multiple Phase Conjugations. *J. Lightwave Technol.* **2016**, *34*, 431–440. [CrossRef]
13. Solis-Trapala, K.; Inoue, T.; Namiki, S. Signal power asymmetry tolerance of an optical phase conjugation- based nonlinear compensation system. In Proceeding of European Conference and Exhibition on Optical Communication (ECOC), (IEEE, 2014), Cannes, France, 21–25 September 2014. paper We.2.5.4.
14. Phillips, I.; Tan, M.; Stephens, M.F.; McCarthy, M.; Giacoumidis, E.; Sygletos, S.; Rosa, P.; Fabbri, S.; Le, S.T.; Kanesan, T.; et al. Exceeding the Nonlinear-Shannon Limit using Raman Laser Based Amplification and Optical Phase Conjugation. In Proceedings of the Optical Fiber Communication Conference, OSA Technical Digest, San Francisco, CA, USA, 9–13 March 2014.
15. Stephens, M.F.C.; Tan, M.; Phillips, I.D.; Sygletos, S.; Harper, P.; Doran, N.J. 1.14Tb/s DP-QPSK WDM polarization-diverse optical phase conjugation. *Opt. Express* **2014**, *22*, 11840. [CrossRef] [PubMed]
16. Da Ros, F.; Yankov, M.P.; Silva, E.P.d.; Pu, M.; Ottaviano, L.; Hu, H.; Semenova, E.; Forchhammer, S.; Zibar, D.; Galili, M.; et al. Characterization and Optimization of a High-Efficiency AlGaAs-On-Insulator-Based Wavelength Convertor for 64- and 256-QAM Signals. *J. Lightwave Technol.* **2017**, *35*, 3750–3757. [CrossRef]
17. Da Ros, F.; Edson, G.; da Silva, E.P.; Peczek, A.; Mai, A.; Petermann, K.; Zimmermann, L.; Oxenløwe, L.K.; Galili, M. Optical Phase Conjugation in a Silicon Waveguide With Lateral p-i-n Diode for Nonlinearity Compensation. *J. Lightwave Technol.* **2019**, *37*, 323–329. [CrossRef]
18. Umeki, T.; Kazama, T.; Sano, A.; Shibahara, K.; Suzuki, K.; Abe, M.; Takenouchi, H.; Miyamoto, Y. Simultaneous nonlinearity mitigation in 92 × 180-Gbit/s PDM-16QAM transmission over 3840 km using PPLN-based guard-band-less optical phase conjugation. *Opt. Express* **2016**, *24*, 16945. [CrossRef]
19. Hu, H.; Jopson, R.M.; Gnauck, A.H.; Randel, S.; Chandrasekhar, S. Fiber nonlinearity mitigation of WDM-PDM QPSK/16-QAM signals using fiber-optic parametric amplifiers based multiple optical phase conjugations. *Opt. Express* **2017**, *25*, 1618. [CrossRef]
20. Namiki, S.; Solis-Trapala, K.; Tan, H.N.; Pelusi, M.; Inoue, T. Multi-Channel Cascadable Parametric Signal Processing for Wavelength Conversion and Nonlinearity Compensation. *J. Lightwave Technol.* **2017**, *35*, 815–823. [CrossRef]

21. Ellis, A.D.; McCarthy, M.E.; Al-Khateeb, M.A.Z.; Sorokina, M.; Doran, N.J. Performance limits in optical communications due to fiber nonlinearity. *Adv. Opt. Photonics* **2019**, *9*, 429–503. [CrossRef]

22. Tan, M.; Al-Khateeb, M.A.Z.; Iqbal, M.A.; Ellis, A.D. Distributed Raman Amplification for Combating Optical Nonlinearities in Fibre Transmission. In Proceeding of 2018 Conference on Lasers and Electro-Optics Pacific Rim (CLEO-PR), Hong Kong, China, 29 July–3 August 2018.

23. Al-Khateeb, M.; Tan, M.; Zhang, T.; Ellis, A. Combating Fiber Nonlinearity Using Dual-Order Raman Amplification and OPC. *IEEE Photonics Technol. Lett.* **2019**, *31*, 877–880. [CrossRef]

24. Bidaki, E.; Kumar, S. A Raman-pumped Dispersion and Nonlinearity Compensating Fiber for Fiber Optic Communications. *IEEE Photonics J.* **2020**, *12*, 720017. [CrossRef]

25. Al-Khateeb, M.A.Z.; Tan, M.; Iqbal, M.A.; Ali, A.; McCarthy, M.E.; Harper, P.; Ellis, A.D. Experimental demonstration of 72% reach enhancement of 3.6Tbps optical transmission system using mid-link optical phase conjugation. *Opt. Express* **2018**, *26*, 23960–23968. [CrossRef] [PubMed]

26. Ellis, A.D.; Tan, M.; Iqbal, M.A.; Al-Khateeb, M.A.Z.; Gordienko, V.; Mondaca, G.S.; Fabbri, S.; Stephens, M.F.C.; McCarthy, M.E.; Perentos, A.; et al. 4 Tb/s Transmission Reach Enhancement Using 10 × 400 Gb/s Super-Channels and Polarization Insensitive Dual Band Optical Phase Conjugation. *J. Lightwave Technol.* **2016**, *34*, 1717–1723. [CrossRef]

27. Al-Khateeb, M.A.Z.; Iqbal, M.A.; Tan, M.; Ali, A.; McCarthy, M.E.; Harper, P.; Ellis, A.D. Analysis of the nonlinear Kerr effects in optical transmission systems that deploy optical phase conjugation. *Opt. Express* **2018**, *26*, 3145–3160. [CrossRef] [PubMed]

28. Tan, M.; Al-Khateeb, M.A.Z.; Zhang, T.T.; Ellis, A.D. Fiber Nonlinearity Compensation Using Erbium-Doped-Fiber-Assisted Dual-Order Raman Amplification. In Proceeding of 2019 Conference on Lasers and Electro-Optics (CLEO), San Jose, CA, USA, 1 May 2019.

29. Rosa, P.; Le, S.T.; Rizzelli, G.; Tan, M.; Harper, P.; Ania-Castañón, J.D. Signal power asymmetry optimisation for optical phase conjugation using Raman amplification. *Opt. Express* **2015**, *23*, 31772–31778. [CrossRef] [PubMed]

30. Bissessur, H. Amplifier technologies for unrepeatered links, submarine transmissions. In Proceedings of the Optical Fiber Communication Conference and Exposition and the National Fiber Optic Engineers Conference (OFC/NFOEC), Anaheim, CA, USA, 17–21 March 2013.

31. Tan, M.; Iqbal, M.A.; Nguyen, T.T.; Rosa, P.; Krzczanowicz, L.; Phillips, I.D.; Harper, P.; Forysiak, W. Raman Amplification Optimization in Short-Reach High Data Rate Coherent Transmission Systems. *Sensors* **2021**, *21*, 6521. [CrossRef]

32. Toyoda, K.; Koizumi, Y.; Omiya, T.; Yoshida, M.; Hirooka, T.; Nakazawa, M. Marked performance improvement of 256QAM transmission using a digital back-propagation method. *Opt. Express* **2012**, *20*, 19821–19851. [CrossRef]

33. Galdino, L.; Tan, M.; Alvarado, A.; Lavery, D.; Rosa, P.; Maher, R.; Ania-Castanon, J.D.; Harper, P.; Makovejs, S.; Thomesn, B.C.; et al. Amplification schemes and multi-channel DBP for unrepeatered transmission. *J. Lightwave Technol.* **2016**, *34*, 2221–2227. [CrossRef]

34. Papernyi, S.B.; Xvnnov, V.B.; Koyano, Y.; Yamamoto, H. Sixth-order cascaded Raman amplification. In Proceedings of the OFC/NFOEC Technical Digest. Optical Fiber Communication Conference, Anaheim, CA, USA, 6 March 2005.

35. Tan, M.; Rosa, P.; Phillips, I.D.; Harper, P. Extended Reach of 116 Gb/s DP-QPSK Transmission using Random DFB Fiber Laser Based Raman Amplification and Bidirectional Second-order Pumping. In Proceedings of the Optical Fiber Communication Conference, OSA Technical Digest, Los Angeles, CA, USA, 22–26 March 2015.

36. Tan, M.; Rosa, P.; Le, S.T.; Iqbal, M.A.; Phillips, I.D.; Harper, P. Transmission performance improvement using random DFB laser based Raman amplification and bidirectional second-order pumping. *Opt. Express* **2016**, *24*, 2215–2221. [CrossRef]

37. Iqbal, M.A.; Tan, M.; Harper, P. Enhanced transmission performance using backward-propagated broadband ase pump. *IEEE Photonics Technol. Lett.* **2018**, *30*, 865–868. [CrossRef]

38. Rosa, P.; Rizzelli, G.; Tan, M.; Harper, P.; Ania-Castañón, J.D. Characterisation of random DFB Raman laser amplifier for WDM transmission. *Opt. Express* **2015**, *23*, 28634–28639. [CrossRef]

39. Tan, M.; Rosa, P.; Iqbal, M.A.; Phillips, I.D.; Nuno, J.; Ania-Castañón, J.D.; Harper, P. RIN mitigation in second-order pumped Raman fibre laser based amplification. In Proceedings of the Asia Communications and Photonics Conference, OSA Technical Digest (Optical Society of America, 2015), Hong Kong, China, 19–23 November 2015. paper AM2E.6.

40. Tan, M.; Rosa, P.; Le, S.T.; Dvoyrin, V.V.; Iqbal, M.A.; Sugavanam, S.; Turitsyn, S.K.; Harper, P. RIN mitigation and transmission performance enhancement with forward broadband pump. *IEEE Photonics Technol. Lett.* **2018**, *30*, 254–257. [CrossRef]

41. Rizzelli, G.; Iqbal, M.A.; Gallazzi, F.; Rosa, P.; Tan, M.; Ania-Castañón, J.D.; Krzczanowicz, L.; Corredera, P.; Phillips, I.; Forysiak, W.; et al. Impact of input FBG reflectivity and forward pump power on RIN transfer in ultralong Raman laser amplifiers. *Opt. Express* **2016**, *24*, 29170–29175. [CrossRef] [PubMed]

42. Tan, M.; Rosa, P.; Phillips, I.D.; Harper, P. Long-haul Transmission Performance Evaluation of Ultra-long Raman Fiber Laser Based Amplification Influenced by Second Order Co-pumping. In Proceeding of Asia Communications and Photonics Conference, OSA Technical Digest (online) (Optical Society of America, 2014), Shanghai, China, 11–14 November 2014. paper ATh1E.4.

43. Tan, M.; Rosa, P.; Le, S.T.; Phillips, I.D.; Harper, P. Evaluation of 100g DP-QPSK long-haul transmission performance using second order co-pumped Raman laser based amplification. *Opt. Express* **2015**, *23*, 22181–22189. [CrossRef] [PubMed]

44. Ania-Castañón, J.D.; Karalekas, V.; Harper, P.; Turitsyn, S.K. Simultaneous spatial and spectral transparency in ultralong fiber lasers. *Phys. Rev. Lett.* **2008**, *101*, 123903. [CrossRef]

45. Hazarika, P.; Tan, M.; Iqbal, M.A.; Krzczanowicz, L.; Ali, A.; Phillips, I.D.; Harper, P.; Forysiak, W. RIN induced penalties in G.654.E and G.652.D based distributed Raman amplifiers for coherent transmission systems. *Opt. Express* **2021**, *29*, 32081–32088. [CrossRef] [PubMed]

46. Cheng, J.; Tang, M.; Lau, A.P.T.; Lu, C.; Wang, L.; Dong, Z.; Bilal, S.B.; Fu, S.; Shum, P.P.; Liu, D. Pump RIN-induced impairments in unrepeatered transmission systems using distributed Raman amplifier. *Opt. Express* **2015**, *23*, 11838–11854. [CrossRef] [PubMed]

47. Rosa, P.; Tan, M.; Le, S.T.; Phillips, I.D.; Ania-Castañón, J.; Sygletos, S.; Harper, P. Unrepeatered DP-QPSK Transmission Over 352.8 km SMF Using Random DFB Fiber Laser Amplification. *IEEE Photonics Technol. Lett.* **2015**, *27*, 1189–1192. [CrossRef]

48. Rosa, P.; Rizzelli, G.; Pang, X.; Ozolins, O.; Udalcovs, A.; Tan, M.; Jaworski, M.; Marciniak, M.; Sergeyev, S.; Schatz, R.; et al. Unrepeatered 240-km 64-QAM Transmission Using Distributed Raman Amplification over SMF Fiber. *Appl. Sci.* **2020**, *10*, 1433. [CrossRef]

49. Bouteiller, J.-C.; Brar, K.; Headley, C. Quasi-constant Signal Power Transmission. In Proceeding of ECOC, Copenhagen, Demark, 8–12 September 2002. paper Symposium 3.4.

50. Bouteiller, J.-C.; Brar, K.; Bromage, J.; Radic, S.; Headley, C. Dual-order Raman pump. *IEEE Photonics Technol. Lett.* **2003**, *15*, 212–214. [CrossRef]

51. Lundberg, L.; Andrekson, P.; Karlsson, M. Power consumption analysis of Hybrid EDFA/Raman amplifiers in long-haul transmission systems. *J. Lightwave Technol.* **2017**, *35*, 2132–2142. [CrossRef]

52. Frauk, M.S.; Savory, S.J. Digital Signal Processing for Coherent Transceivers Employing Multilevel Formats. *J. Lightwave Technol.* **2017**, *35*, 1125–1141. [CrossRef]

53. Zhang, J.; Li, X.; Dong, Z. Digital Nonlinear Compensation Based on the Modified Logarithmic Step Size. *J. Lightwave Technol.* **2013**, *31*, 3546–3555. [CrossRef]

54. Nguyen, T.T.; Zhang, T.; Giacoumidis, E.; Ali, A.; Tan, M.; Harper, P.; Barry, L.; Ellis, A. Coupled Transceiver-Fiber Nonlinearity Compensation Based on Machine Learning for Probabilistic Shaping System. *J. Lightwave Technol.* **2021**, *39*, 388–399. [CrossRef]

55. Ellis, A.D.; McCarthy, M.E.; Al-Khateeb, M.A.Z.; Sygletos, S. Capacity limits of systems employing multiple optical phase conjugators. *Opt. Express* **2015**, *23*, 20381–20393. [CrossRef] [PubMed]

Review

A Review: Point Cloud-Based 3D Human Joints Estimation

Tianxu Xu [1,†]**, Dong An** [1,†]**, Yuetong Jia** [1] **and Yang Yue** [1,2,*]

[1] Institute of Modern Optics, Nankai University, Tianjin 300350, China; 1120180105@mail.nankai.edu.cn (T.X.); 1120190109@mail.nankai.edu.cn (D.A.); 1911343@mail.nankai.edu.cn (Y.J.)

[2] Angle AI (Tianjin) Technology Company Ltd., Tianjin 300450, China

[*] Correspondence: yueyang@nankai.edu.cn

[†] These authors contributed equally to this work.

Abstract: Joint estimation of the human body is suitable for many fields such as human–computer interaction, autonomous driving, video analysis and virtual reality. Although many depth-based researches have been classified and generalized in previous review or survey papers, the point cloud-based pose estimation of human body is still difficult due to the disorder and rotation invariance of the point cloud. In this review, we summarize the recent development on the point cloud-based pose estimation of the human body. The existing works are divided into three categories based on their working principles, including template-based method, feature-based method and machine learning-based method. Especially, the significant works are highlighted with a detailed introduction to analyze their characteristics and limitations. The widely used datasets in the field are summarized, and quantitative comparisons are provided for the representative methods. Moreover, this review helps further understand the pertinent applications in many frontier research directions. Finally, we conclude the challenges involved and problems to be solved in future researches.

Keywords: point cloud; joint estimation; skeleton extraction; depth sensor; skeleton tracking; computer vision; human representation; convolutional neural network; random tree walk; random forest; geodesic features; global features; deformation model; hand pose tracking; action recognition

Citation: Xu, T.; An, D.; Jia, Y.; Yue, Y. A Review: Point Cloud-Based 3D Human Joints Estimation. *Sensors* **2021**, *21*, 1684. https://doi.org/10.3390/s21051684

Academic Editor: Roberto Vezzani

Received: 28 January 2021
Accepted: 23 February 2021
Published: 1 March 2021

Publisher's Note: MDPI stays neutral with regard to jurisdictional claims in published maps and institutional affiliations.

1. Introduction

The depth camera can provide the ranging information from a single depth image or a point cloud for a variety of applications, such as gaming, three-dimensional (3D) reconstruction and object recognition. Many human-centered tasks based on depth camera had been investigated in the last few years, as shown in Figure 1. For example, 3D human reconstruction is the process of recovering a 3D human surface model by finding the accurate correspondence between frames [1,2]. The 3D segmentation technology of human body is the most critical technology in applications such as digital clothing and computer animation [3]. The health monitoring system using the depth information can check the diseased parts of the human body to facilitate the guidance of rehabilitation training [4]. The size measurement of human body based on the depth camera is a safe and non-contact fast measurement method, which overcome the challenges of high cost and bulky electronic scanners [5]. Human behavior recognition, as a fundamental research problem, is an extremely significant component and extensively studied research subject in computer vision [6]. The ultimate objective of encoding human body is to extract the various joints of a predefined skeleton in a simplified manner.

Conventional methods for detecting joints of human body utilize two-dimensional (2D) images or video, taken by traditional cameras. Significant progress has been made for these methods in recent years, by leveraging the powerful deep learning. However, there are still some limitations on human pose estimation using only 2D images, due to the coexisting complex backgrounds, variable viewpoints, highly flexible poses, etc. Additional depth information can provide enriched 3D data to overcome the limitation of 2D data.

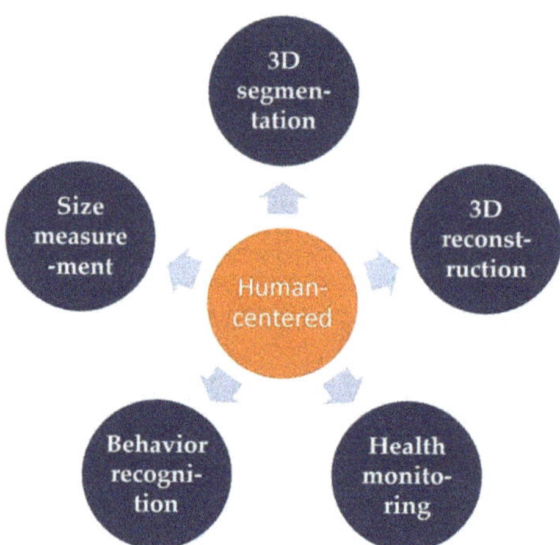

Figure 1. Human-centered applications of depth sensor.

The purpose of depth camera-based 3D human pose estimation is to locate the (x, y, z) coordinates of joints in 3D space. Ideally, once the captured human pose changes, the joints can still be reliably estimated. Figure 2 describes the specific overview of the 3D joints extraction. The first step of this process is to capture the human poses by a depth sensor. Since the obtained poses of human body contain redundant information, it is necessary to process the data in advance. Next, the pre-processed data is used to calculate the 3D coordinates of the joints using special methods, such as template-based method, feature-based method, and machine learning-based method. Finally, the error analysis is performed.

Usually 3D data formats include depth map, point cloud, mesh, and voxel grid, etc. Here a point cloud is a collection of points in the 3D space. In addition to the commonly used 3D coordinate information, point cloud can also carry other dimensional information such as color and normal vector. The point cloud can be obtained directly through the depth sensors. Compared with the voxel grid, the storage space of the point cloud is smaller, and the geometric information can still be expressed well after the rotation. Compared with the mesh method, the point cloud is easily obtained, while there is not direct method to acquire the mesh data. Compared with the depth map, the point cloud represents the 3D object in a more intuitive way. Moreover, the conversion between the point cloud and the other 3D formats is quite straight forward. The widely used open source libraries for 3D point cloud processing are mainly Point Cloud Library (PCL) [7] and Open3D [8]. PCL is a cross-platform C++ library, which implements a large number of point cloud-related general algorithms and efficient data structures, involving point cloud acquisition, filtering, segmentation, registration, retrieval, feature extraction, recognition, tracking, surface reconstruction, visualization, etc. It can support multiple operating systems such as Windows, Linux, Android, Mac OS X, and some embedded real-time systems. Open3D is a modern library that can support the rapid software development for 3D data processing. A set of data structures and algorithms are exposed in C++ and Python, its core features include 3D data structure, 3D data processing algorithms, scene reconstruction, and 3D visualization. PCL is more mature with a large number of data structures and algorithms for 3D data processing. In contrast, Open3D can be installed and used in the python environment, and the programming is faster and simpler.

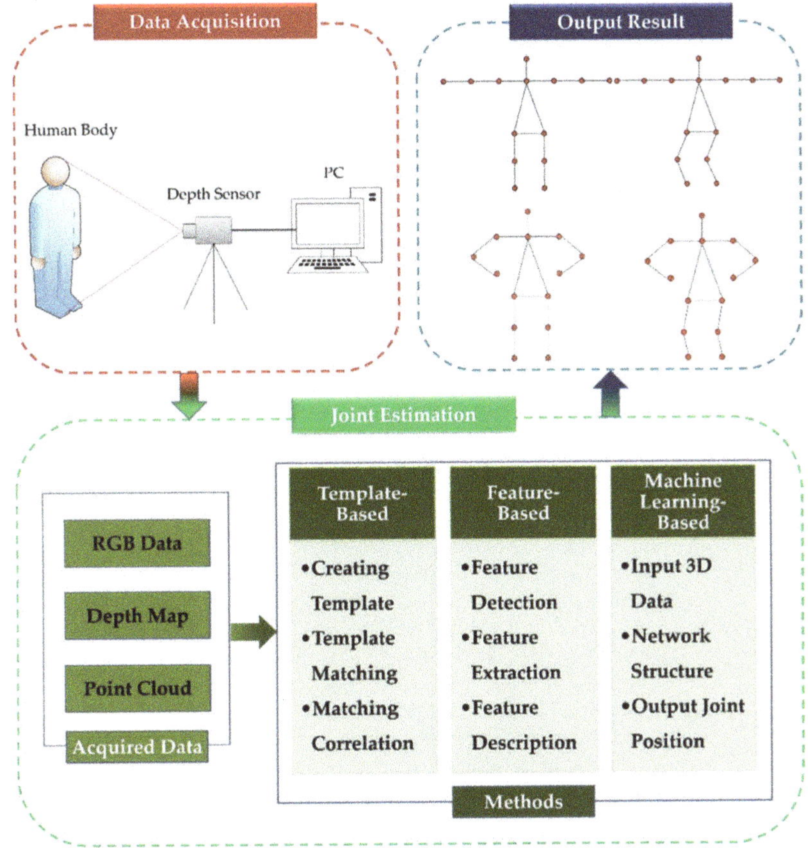

Figure 2. The flow diagram of point cloud-based 3D joints extraction.

The aim of this survey is to provide a comprehensive overview of 3D human joints extraction based on point cloud. The survey mainly focuses on publication of point cloud-based joint estimation for human body in computer vision. Here, the point cloud of human body is directly captured from 3D ranging devices. The involved content mainly includes the datasets of relevant pose, the methods of point cloud-based joint extraction, and the applications of point cloud-based joint extraction. The summarized methods are shown in Figure 3, including template-based methods, feature-based methods and machine learning-based methods. Compared with the existing surveys, the main contributions of this review include:

1. To the best of our knowledge, this is the first review to summarize the point cloud-based 3D joints estimation of human body, thereby providing readers with a complete overview of the latest researches and developments in the field.
2. The review categorizes the advanced methods based on their working principles in a comprehensive way, and we enumerate some milestone works in recent years.
3. The datasets and applications of point cloud-based 3D joint extraction are analyzed. In addition, the results from different literatures are summarized and compared.

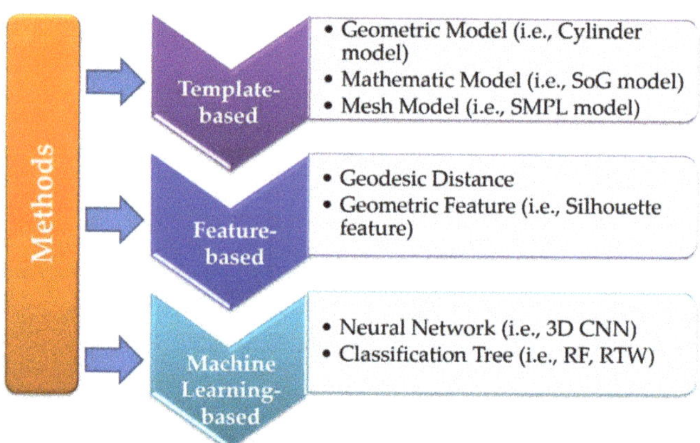

Figure 3. Categorization of the methods for the 3D joints extraction based on point cloud. SoG is sums of spatial gaussians; SMPL is skinned multi-person linear; CNN is convolutional neural network; RF is random forest; RTW is random tree walk.

The remainder of this review is organized as following: Section 2 introduces sets of devices for range detection; Section 3 reviews the existing point cloud-based methods of human joints extraction; Section 4 enumerates the 3D human dataset; Section 5 discusses various applications of point cloud-based joint extraction; Section 6 concludes this survey with potential research directions in the future.

2. Depth Sensors for 3D Data Acquisition

According to the ranging principle, we divide the depth sensors into three categories as shown in Figure 4, binocular stereo vision, time-of-flight (ToF) and structured light technologies. Early research focused on the passive method, such as binocular stereo vision, to calculate the depth information. Typically, two cameras were used to take pictures of the object from different perspectives. This mechanism is similar to imaging by two eyes of human. However, the result can be easily affected by the texture of object, and it is also time-consuming in the registration process. Compared with the passive ranging method, rapid active ranging shows obvious advantages. On one hand, radar and Lidar are commonly used in military to reconnoiter and detect battlefields in various environments, and it is also manipulated for obstacle detection in automatic driving technology. On the other hand, with the commercial application of low-cost depth cameras, the related research based on depth cameras has gradually unlocked new applications in the mobile and intelligent terminal devices.

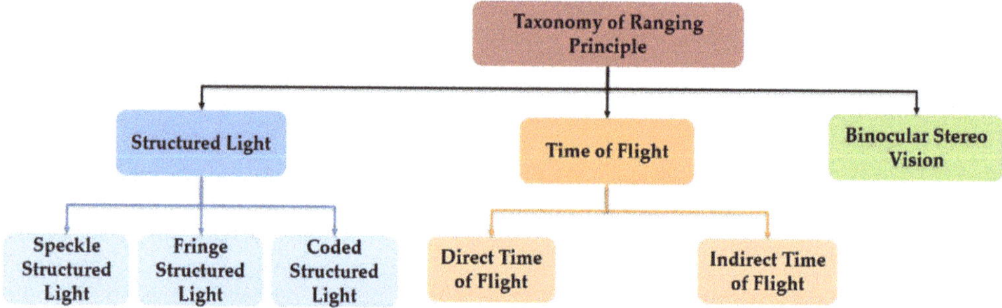

Figure 4. Classification of depth sensors for 3D data acquisition.

The ToF depth camera first continuously sends light pulses to the detected object, and then the sensor is used to receive the light returned from the object. The final distance is calculated when the flight (round trip) time of the detected light pulse is obtained. ToF sensors are divided into two types: direct time-of-flight (dToF) and indirect time-of-flight (iToF) sensors according to different modulation methods. Lighting units of dToF generally use LEDs or lasers, including laser diodes and vertical cavity surface emitting lasers (VCSELs), to emit high-performance pulsed light, which directly measures the time difference from the emitter to the receiver and multiplies it by the speed of light to measure the relative distance of the object. The receiver must be a special sensor with very high accuracy, so it is difficult to reduce the cost and miniaturization. The light emitted by the iToF sensor is modulated by a continuous wave, whose intensity changes regularly. According to the selection of detected distance, it can be divided into pulsed light and continuous light. Next, the iToF depth cameras compare the signal difference between the emitted signal and the reflected signal, and then multiply it by the speed of light to get the relative distance. In addition, the response speed of the iToF receiver is not as fast as dToF, and it cannot accurately sense for the sub-nanosecond time difference. Fundamentally, radar and Lidar are also a kind of ToF sensor. The former uses millimeter waves, and has stronger anti-interference ability, the latter emits laser signals, and has higher detection accuracy. Structured light cameras project the invisible pseudo-random light spots to the detected object through the infrared (IR) emitter. According to the produced different light spots, they can be divided into speckle structured light, fringe structured light, and coded structured light cameras. The projected light spot is unique and known in the spatial distribution, and have been pre-stored in the structured light memory. The size and shape of these speckles projected on the observed object vary according to the distance and direction of the object and the camera. The captured spots are compared with the known spots, and then the depth information is obtained. Different depth cameras can be selected according to specific parameters, and the detailed comparison is shown in Table 1.

Table 1. Comparison of different depth cameras.

Depth Cameras	Binocular Stereo Vision	ToF	Structured Light
	low hardware requirements	long detection distance	convenience for miniaturization
Advantages	low cost	large tolerance to ambient range	low resource consumption
	high robustness to light disturbance	high frame rate	high resolution
	large calculation complexity	high equipment requirements	small tolerance to ambient light
Disadvantages	strong object texture dependence	high resource consumption	short detection range
	limited measurement range	low edge accuracy	high noise
Representative	ZED 2K Stereo Camera	Kinect v2	Kinect v1
	BumbleBee	Intel RealSense L515	XTion

3. Methods of Point Cloud-Based Joint Estimation

This section mainly describes several methods to achieve human joint extraction. Existing surveys have made qualitative comparisons of the joint extraction techniques from the perspective of RGB or depth map. We limit our attention to point cloud-based methods, which are mainly divided into feature-based methods, template-based methods and machine learning-based methods. Each method is discussed in each section, respectively.

3.1. Template-Based Methods

The human body is a flexible and complex object with many specific features, such as movement structure, body shape, surface texture, body parts or joint positions. A mature human model is not necessary to contain all the human attributes. Otherwise, it should meet the specific task of combining and describing human poses. The template-based method is intuitive and simple. It judges the motion category by comparing the similarity between the detected object and the constructed template. The existing template-based algorithms can be roughly divided into three categories according to their principles, including geometric model, mathematical model, and mesh model. When the model is selected to match with the observed point cloud of the human body, the joints are regarded as the connection of the rigid part to achieve pose estimation. A complex model usually features more characteristic parameters; it can provide a better approximation for the human body to achieve improved reality and accuracy.

3.1.1. Geometric Model

In general, a geometric model roughly divides the human body into several parts. Each part can be regarded as rigid and then fitted by the 3D geometric shapes, including general cylinder, ellipse, and rectangle. Knoop et al. [9] proposed a new method of fusing different input signals for human tracking. The algorithm can process 2D and 3D input data from different sensors (such as ToF camera, stereo or single-ocular images). For the tracking system, a 3D human model was built with several parts, in which each part was represented by a degenerate cylinder. The top view and bottom view of each cylinder can be regarded as an ellipse structure. Moreover, the two ellipses cannot rotate with each other and their planes were parallel. Therefore, a cylinder model needed to be described by five parameters: the major axes and minor axes of these two ellipses, together with the length of the cylinder. The entire 3D human model was composed of 10 cylinders, of which the torso started to extend as the root node. Each child node was described by a degenerate cylinder and the corresponding transformation of its parent node.

The Head–Neck–Trunk (HNT) deformable template, represented by circles, trapezoids, rectangles, and trapezoids, was proposed in 2011 [10]. Once the HNT template works, the limbs (i.e., two arms and two legs) would be detected and fitted with rectangles. The end points of the rectangle were regarded as the joints of the human body, and its depth information was used to determine whether the human body was in a self-occlusion state. When self-occlusion occurred, the part segmentation of the human body was triggered, and then the segmented limbs were fitted separately. Inversely, the joints of the human body were directly obtained containing the contact points between the geometric shapes and the end points. Suau et al. [11] proposed a fast method to localize five joints of human body on the point cloud. In this method, the geometric deformation model established by the basic curve evolution theory and the level set method [12] was adopted to spread the topological structure of the human body. Additionally, the Narrow Band Level Set (NBLS) method [13] was also expanded to filter the 2.5D data according to its physical area. With the purpose of maintaining the connectivity on the depth surface to facilitate the extraction of topological features, the calculated NBLS map was filled, and finally the geodesic distance was used to quickly locate the five end points corresponding the five extreme joints. Lehment et al. [14] used an ellipsoid model for the upper-body with nine basic body modules, including left/right upper arms, lower arms, hands, head, torso and neck. Since the ellipsoid is a 3D equivalent ellipse, it is easy to be generated and controlled. Even in the case of having no clue about the color or texture, it can also find the nearest neighbor points with the input point cloud to calculate the similarity of the likelihood function.

Unlike the above method, Sigalas et al. [15] used 2D information to estimate the pose of the 3D torso. The initial face identification of the human in 2D image was used to segment the area of human body from the background. Based on the illumination, scale and pose invariant features on the 2D silhouette, the 2D silhouette was extracted from a 2D body, and then the curve analysis was performed to initially assume the area of the human shoulder. Meanwhile, the 3D information of the point cloud was meshed to estimate the 3D shoulder coordinates. The ellipsoid model was finally fitted with the torso area of the human body by least-squares optimization; a set of anthropometric standards were also applied to further refine the 3D torso pose.

In order to increase the robustness of the algorithm, Sigalas et al. [16] further demonstrated a multi-person tracking system, as shown in Figure 5, in which human segmentation and pose tracking were contained. Human segmentation detected multiple human bodies by face detection in the depth map, and each human body was segmented individually. In the meantime, the length information of each part was calculated. Pose tracking first defined a body model with a head, upper and lower torsos, arms, and legs. Ellipse and circular were used to represent the upper and lower torsos, while cylinder could be implemented to fit for the remaining parts. Additionally, each part had a length limit. The point cloud of the human body obtained by the depth camera was rotated to the top view, and the reprojection ratio f_{reproj} in Equation (1) was introduced as a matching index.

$$f_{reproj} = \frac{N_{Pr}}{N_{3D}} \tag{1}$$

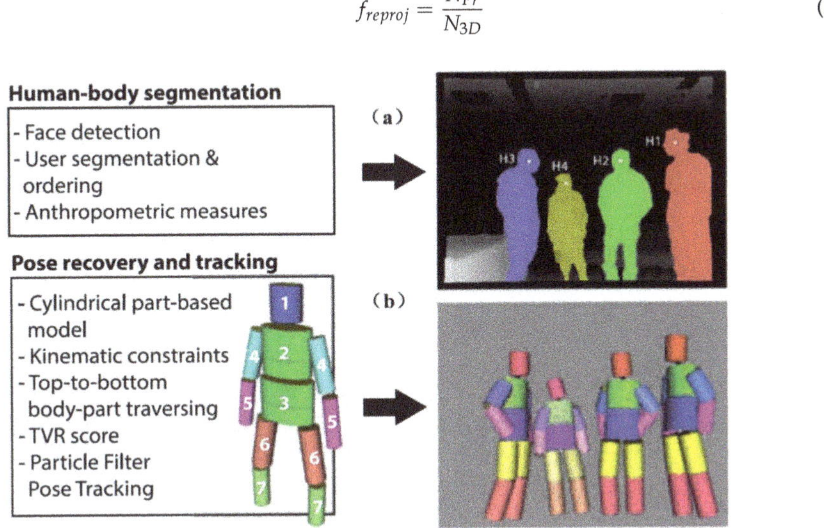

Figure 5. A geometric model can be applied to human joint estimation. (**a**) includes the human body segmentation and depth-based ordering and (**b**) includes the pose recovery and tracking. Figure from [16].

Multiple views by rotating the cylinder model around the *x*-axis can be generated, including occluded and non-occluded. For each view of the cylinder, the corresponding reprojection ratio visible points N_{Pr} to the total number of 3D points N_{3D} in the point cloud was calculated. Its value varied with the view, and it reached the minimum in the top view of the cylinder when the view-axis of the camera was aligned with the long axis of the cylinder.

Based on T-pose, Wu et al. [17] created a simplified human skeleton model with customized parameters to adapt to different body types. The depth image and the corresponding 3D point cloud, as a pair of inputs, were first pre-processed and initialized to obtain personalized parameters. The torso part could be detected on the binarized image, the centroid of this part was calculated as the root node afterwards, and then using the root node as the parent node to iteratively find other child nodes. After obtaining the length between nodes, the human skeleton information was obtained by matching with skeleton model and further optimizing the joint angle. Besides, this method used the threshold segment to solve the self-occlusion problem.

3.1.2. Mathematical Model

Mathematical model mainly transfers conceptual knowledge commonly used in mathematics to model construction. The basic idea is to build a model with the representation method of probability distribution to list each possible result and give their probabilities. Significant amount of work has been accomplished using Gaussian Mixture Model (GMM) in recent years. GMM is to establish a mixed model based on multiple Gaussian distributions for each pixel in the image. The parameters in the model are continuously updated according to the observed image, and background estimation is performed at the same time. Based on GMM, an algorithm [18] was proposed based on a single depth camera to estimate the pose and shape of the human body in real time. Due to the probabilistic measurement, it did not require explicit point correspondences. The articulated deformation model, which is based on exponential-maps, can direct embed into the GMM model. However, this algorithm simply used the first few frames to acquire human pose in the dynamic scenes, which usually did not provide the complete information. To cope with the time-varying articulated human body shape, Xu et al. [19] applied a GMM model to establish the pose and shape of the observed user. This method obtained the correspondence between the model and the user, and realized the shape estimation of human body based on multiple RGB-D sensors without any priori information. Compared with a single view case, depth data from multiple RGB-D sensors can not only handle more complex poses, especially occlusion situations, but also can be used to achieved different types of shape estimation by changing body attributes such as height, weight or other physical characteristics. Ge et al. [20] constructed a new non-rigid joints registration framework for human pose estimation by improving the two latest registration techniques. One is Coherent Point Drift (CPD), and the other is Articulated Iterative Closest Point (AICP). The GMM model was applied to initialize the standard pose of the human body through the CPD, and then AICP was employed with other pose point clouds to complete the pose estimation task. In the follow-up work, for incomplete data caused by self-occlusion and view changes, an effective pose tracking strategy was introduced to process continuous depth data [21,22]. Each new frame initialized a new template, which effectively reduced the ambiguity and uncertainty in the process of visible point extraction.

Stoll et al. [23] proposed Sums of spatial Gaussians (SoG) in 2015, which used a quadtree to gather image pixels with similar color values into a larger square. It demonstrated remarkable performance for 2D data. Each square was represented by a Gaussian function, and then a set of isotropic Gaussian components constituted the SoG. Inspired by SoG, Ding et al. [24] presented Generalized SoG (G-SoG), which used an anisotropic Gaussian function with less calculation to represent the entire human body. On this basis, they expanded the 3D express form of SoG by grouping 3D parts of the point cloud with similar depth into voxels. The 3D Gaussian model only contained spatial statistical data, but not color information.

Both SoG and G-SoG involve pose tracking of different characters. The former represents observed point cloud through effective octree division, and the latter embeds a quaternion-based articulated skeleton to create a standard human template model. A single un-normalized 3D Gaussian G can be expressed as Equation (2):

$$G(x) = \exp\left(-\frac{\|x - \mu^2\|}{2\sigma^2}\right),$$ (2)

where x is 3D coordinates, μ and σ^2 are the mean and the variance, respectively. SoG has the form as Equation (3),

$$K(x) = \sum_{i=1}^{n} G_i(x)$$ (3)

A 3×3 covariance matrix is introduced into Equation (2) to replace the variance σ^2, an anisotropic Gaussian in Equation (4) is obtained as,

$$G(x) = exp(-\frac{1}{2}(x-\mu)^T \begin{bmatrix} C_{11} & C_{12} & C_{13} \\ C_{12} & C_{22} & C_{23} \\ C_{13} & C_{23} & C_{33} \end{bmatrix} (x-\mu)) \tag{4}$$

SoG was used to represent the point cloud of the human body, and then registered it with the G-SoG body template for human tracking. An Energy function in Equation (5), including similarity term, continuity term and visibility term, to describe the similarity between G-SoG and SoG [25],

$$\hat{\theta} = \underset{\theta}{arg\min} \sum_{i \in K_m} -E_{sim}^{(i)}(\theta) \cdot Vis(i) + \lambda_{con} E_{con}(\theta), \tag{5}$$

the first term emphasizes the similarity of the two models, V_{is} gives the visible state of each Gaussian function, and the second term is added to smooth the pose estimation. In addition, the overlaps on the 2D projection plane of the Gaussian functions were used to judge whether there is occlusion.

Based on the previous work, the author expanded the previous framework and proposed an articulated and generalized Gaussian kernel correlation (GKC)-based system [26], as shown in Figure 6, which supported subject-specific shape modeling and articulated pose estimation for the whole body and hands.

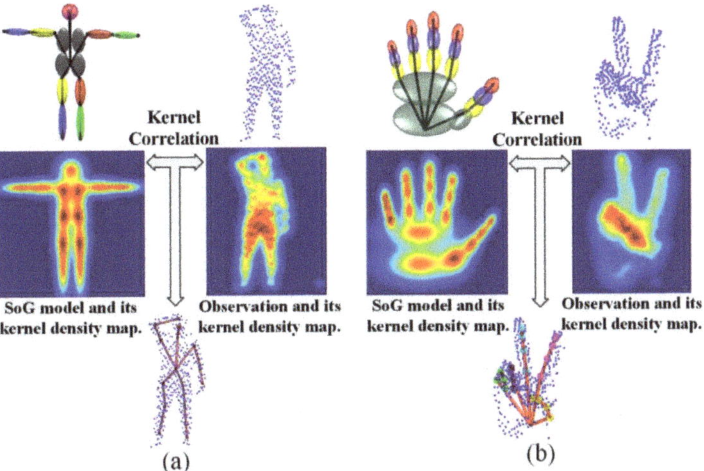

Figure 6. A conceptual scheme of mathematical model-based human joint estimation. Articulated pose estimation for the full body (**a**) and hand (**b**). The 1st row shows the Sums of spatial Gaussians (SoG) -based template models and an observed point cloud. Their corresponding Gaussian kernel density maps are depicted in the 2nd row, followed by the pose estimation results in the 3rd row. Figure from [26].

Apart from the method based on Gaussian distribution, Ganapathi et al. advanced a real-time tracking algorithm based on Maximum A Posteriori (MAP) inference in a probabilistic temporal model, and the human pose of each part was updated with Iterative Closest Point (ICP) algorithm. A two-stage method was proposed to solve the problem of recovering the human pose from a single depth map [27]. In the first stage, course template found in a large model dataset was used to make skeleton deformation, and then in the second stage, the detailed part of the human shape was restored by Stitched Puppet model [28] to fit the deformed model.

3.1.3. Mesh Model

The mesh model is composed of many small polygonal patches in the computer to form the surface in the real world. Through the parameterized human body model, the structure has specific outer surfaces in addition to the skeleton, which reflects 3D appearance of the human body. Specific details such as characteristics are convenient to judge whether self-occlusion behavior occurs.

Ye et al. [29] built a fast pose detection system. After segmenting and denoising, the human point cloud was aligned with a series of mesh model, and the invisible parts were filled in the alignment process. Next, the shape and pose were deformed to perform fine-tuning. For the point cloud that failed to be registered, they searched the best alignment model again to complete the joint extraction process. Grest et al. [30] used the ICP algorithm with nonlinear optimization technology to achieve the purpose of aligning the mesh model and the human point cloud. By using the ICP algorithm, Park et al. [31] also recorded and processed multiple depth point clouds of single person from different perspectives to capture the shape of the entire body. Using template matching and Principal Component Analysis (PCA), a statistical body model representing a variety of human shapes and poses can be generated. PCA shortened the searching time by projecting the data into a low-dimensional Principal Component (PC) space. The ICP algorithm was adopted to fit the subject-specific human body model and depth data frame by frame, so that the accuracy of the original joint positions estimated by the Software Development Kit (SDK) was improved.

Hesse et al. [32] exerted a combination of texture model and random forest to classify body parts. According to the parts of human body, the position of the human joint was estimated. Vasileiadis et al. [33] used 3D Signed Distance Functions (SDF) data to represent the model, which was extended by a supplementary mechanism to track the pose of the human body in the depth sequence. In the actual multi-person interaction scene, the depth data of the human body in different perspectives was collected [34], and the mesh model was used for fitting to eliminate contact joint error. A new unsupervised framework was proposed to eliminate the influence of noise [35]. The method consisted of three steps: the deformed model and the human point cloud were registered with non-rigid point method to establish point correspondence, and the skeleton structure was extracted from the new point set sequence based on the cluster. Finally, Linear Blend Skinning (LBS)-based joint learning refine the positions. Huang et al. [36] estimated the joint positions by fitting a reference surface model, which included a reference triangle mesh surface and an inherent tree-shaped skeleton. Walsman et al. [37] utilized mesh templates to track human pose in real time and reconstructed high-resolution surface silhouette, so that it can facilitate gesture recognition and motion prediction using commercial depth sensors and GPU hardware.

Among all the mesh models, as a prominent and parametric human body model, Skinned Multi-Person Linear (SMPL) model can carry out arbitrary shape model and animation drive. This method can simulate the bulges and depressions of human muscles during limb movement. Therefore, the surface distortion of the human body during exercise can be avoided, and the appearance of human muscle stretching, and contraction can be accurately described. Zhou et al. [38] employed MobileNet to build a 2D human skeleton model, which facilitated the initialization of the point cloud. And then the customized SMPL model was fitted to the observed point cloud. The error was gradually reduced between the SMPL model and the actual observed point cloud by minimizing the loss function.

In order to enhance the generalization ability of the model, Joo et al. [39] designed a unified deformable model "Frank" to capture human motion at multiple scales without markers, including facial expressions, body motions and gestures. Figure 7 illustrates the main components of Frank. Each part is fitted with FaceWarehouse [40], SMPL model and artist-defined hand, respectively. Finally, the three models are partially spliced to capture human body motion and subtle facial expressions. The seamless mesh V^U in Equation (6) was denoted as the motion and shape of the target subject. The main components of the Frank model M^U include motion factor θ^U, shape factor φ^U, and global transformation factor t^U.

$$V^U = M^U\left(\theta^U + \varphi^U + t^U\right) \tag{6}$$

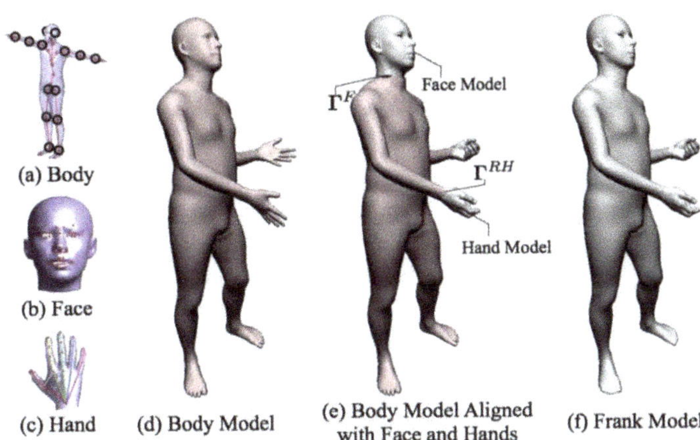

Figure 7. A mesh model is used to fit with point cloud of human body. (**a**) SMPL model; (**b**) Face-Warehouse model; and (**c**) artist-defined hand model. In (**a–c**), the red dots represent 3D positions of the corresponding keypoints reconstructed by detectors; (**d**) Body model; (**e**) Face and hand models are aligned with the corresponding parts of the body model; and (**f**) The whole Frank model. Figure from [39].

The motion parameters θ^U in Equation (7) and shape parameters φ^U in Equation (8) of the Frank model are the combination of each sub-model parameters,

$$\theta^U = \left\{\theta^B,\ \theta^F,\ \theta^{LH},\ \theta^{RH}\right\}, \tag{7}$$

$$\varphi^U = \left\{\varphi^B,\ \varphi^F,\ \varphi^{LH},\ \varphi^{RH}\right\}, \tag{8}$$

where B represents SMPL model, F belongs to face model, LH and RH are abbreviations of the left- and right-hand models, respectively. The motion parameter θ^U mainly expresses the overall motion pose of the human body, including the relative angle information of the joints, and the shape parameter φ^U is defined as the ratio of length, width and height.

The next step is to merge the model with the point cloud. There are two cases: one is when the corresponding point can be found between the model and the point cloud, the other is when the corresponding point is not obvious. In the first case, 2D detection first was operated to find the corresponding keypoints in each sub-region, and then converted them to 3D space. In the second case, the ICP algorithm was exerted to register the point cloud with model. The final objective function can be written as Equation (9):

$$E\left(\theta^U, \varphi^U, t^U\right) = E_{keypoints} + E_{icp} + E_{seam} + E_{prior} \tag{9}$$

$E_{keypoints}$ means 3D keypoints detections, the term E_{icp} expresses the cost of ICP algorithm. The skeleton hierarchy of the Frank model was closely connected. However, the independent surface parameterizations of some sub-model may lead to the introduction of discontinuities at the boundary. To avoid this artifact, the difference E_{seam} between the vertices of the last two circles around the seam was minimize. Because the SMPL and Face-Warehouse model did not capture hair and clothes, the full body could not be explained well by the model. This resulted in incorrect registration during ICP. Hence, E_{prior} was set on the model parameters to avoid overfitting the model to these noise sources. Furthermore, a new model, Adam, was derived to better capture the rough geometry of the human body with clothes and hair to match the geometry of the source data more accurately.

This method above showed the potential that the unmarked motion being captured can eventually surpass the mark-based one. The marker-based method is very susceptible to occlusion, which makes it difficult to capture the details of the body and hands at the same time. This work can not only solve the occlusion problem, but also achieve higher precision model fitting results.

3.2. Feature-Based Methods

Global feature is a common feature-based method, which refers to the overall attributes. Common features include color, texture, and shape features. Because it is a low-level visual feature at the pixel level, the global feature possesses low variance, simple calculation, intuitive representation, etc. Among the global features, geodesic distance and geometric feature are commonly used in the point cloud-based applications.

3.2.1. Geodesic Distance

Geodesic distance is literally inferred to be the shortest path distance between two points, which is different from the Euclidean distance that usually being used in the geometrical space. Euclidean distance is the shortest distance between two points in space, and geodetic distance is the shortest path of two points along the surface of the object. To solve the shortest path problem, Dijkstra's algorithm [41] was commonly used, which was a greedy algorithm. This was because after specifying the starting point and ending point, the algorithm always tried to access the next node that is closest to the starting point in each step of the loop, thereby gradually obtaining the shortest distance between the two points.

Krejov et al. [42] located and separated the left and right hands according to the image domain, and then processed each hand in parallel to build a weighted graph on the surface. An effective Dijkstra's algorithm is utilized to traverse the entire graph to find N candidate fingertips. With the shortest path algorithm, multi-touch interaction among multiple users is realized. Phan et al. [43] proposed an online multi-view voting scheme (MVS) running at an interactive rate. It combined the measurement results from multiple sources to generate a fine geodesic distance graph (GDG), and then five geodesic extremes in the GDG were marked as the head, hands, and feet. Assuming that the length of each bone is determined in advance, so additional landmarks are obtained by calculating the centroid of each region, corresponding to the secondary joints of the wrist, elbow, knee, ankle, and neck. To overcome the errors caused by misdetection and occlusion, an improved method using feature point trajectories to correct the error detection was designed [44]. Five extreme points were detected by geodesic distance method. A shoulders template was applied to search for the position of the shoulders. Once the shoulder joint is determined, the geometric midpoint was regarded as the position of the elbow joints. An iterative search method was used to find the elbow point by minimizing the total geodesic distance from the shoulder point to the hand point through the elbow point. Besides, a minimum distance constraint was imposed afterward in the corresponding recognition to predict its most likely spatial position in the next frame for tracking the trajectory of each joint. To solve the problem of detecting and identifying body parts in the depth data at the video frame rate, a solution was proposed to obtain a new interest point detector on the point cloud data [45]. First, the extreme points were detected by using the geodesic distance, and were further divided into hands, feet, or head using local shape descriptors, and 3D direction vector of each point is given. To speed up the search process of candidate points in the human body, a quadtree-based method was utilized to effectively group adjacent data, and then Dijkstra's algorithm was applied on this basis to obtain the feature points. In the tracking process, a noise removal and restoration method based on Kalman filter was used to correct and predict the extreme positions [46].

Combining multiple methods can also provide better accuracy in estimating the position and direction of the joints. Handrich et al. [47] replaced depth information with more complex features describing local geodesic neighborhoods, and then a random forest classifier was used to learn the correct body part from these descriptors. Baak et al. [48] employed the geodesic distance, which was extracted from the input data as a sparse feature to retrieve the pose from a large 3D pose dataset, and merged it with the previous pose to achieve pose tracking. Mohsin et al. [49] described a system for successfully locating specific body parts. Multiple depth sensors were used to collect point clouds from different perspectives to help solve the occlusion problem. In order to locate prominent human limbs, a triangular mesh model was applied to the 3D point cloud, and the ends of the limbs were marked with geodesic distance.

In general, the geodesic distance can only be used to detect the five extreme points of the human body, namely, the head, hands, and feet. A hybrid framework using depth camera to automatically detect joints was proposed [50]. This method divided the joints into two types: implicit joints and dominant joints. Dominant joints include extreme points, elbows, and knees. Implicit joints are points on the trunk, such as the neck and shoulders. The specific extraction process is shown in Figure 8a. Due to the rigidity of the human limbs, the dominant joints are easier to be detected than the implicit ones. First, the geodesic features of the human body are used to establish extreme points.

$$D_g\left(p_0, P\left(x_p, y_p\right)\right) = \sum D_g\left(P\left(x_p, y_p\right), P\left(x_q, y_q\right)\right) \tag{10}$$

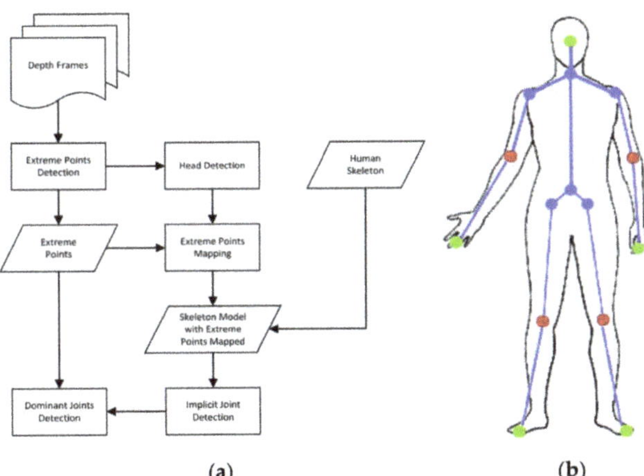

Figure 8. Geodesic distance is used to locate the positions of extreme points. (**a**) describes the overview of the workflow of the proposed method. (**b**) shows the skeleton model used in our method. The green dots represent the extreme points. Blue dots represent implicit joints (neck, waist, shoulders, and hips). Red dots represent dominant joints (elbows and knees). Figure from [50].

In Equation (10), P denotes the point cloud, p_0 is starting point, and $D_g\left(\cdot\right)$ represents the geodesic distance between two random points. If the corresponding relationship between the extreme points and the skeleton model is not given, it is difficult to detect the position of the joints. Therefore, starting from mapping an extreme point to the head, the feature of the area around each extreme point is used to compare with the head model. Each extreme point is gradually mapped to the corresponding part of the human body model. In the skeletal model as described in Figure 8b, the geodesic distance between the head and the hand is smaller than the geodesic distance between the head and the foot, which is the criterion used to separate the hand and the foot joints.

With the above restrictions, the extreme points are found. The human skeleton model is then used to define implicit joints. Assuming that the geodesic distance between the left hand and the left shoulder is shorter than the geodesic distance between the left hand and the right shoulder. The relationship between the left hand and the right hand can be described as Equations (11) and (12):

$$D_g(p_{Lh}, j_{Ls}) < D_g(p_{Lh}, j_{Rs}),$$ (11)

$$D_g(p_{Rh}, j_{Rs}) < D_g(p_{Rh}, j_{Ls})$$ (12)

By adding constraints such as Euler angle and geodesic distance ratio, the joint candidates were ensured to show the degree of curvature of the path. The strategy based on the global shortest path was adopted to detect the dominant joint candidates, such as elbow and knee joint, and then the shortest paths for specific detection were further used to locate these joint. Furthermore, to deal with self-occlusion, when the distance map is updated, the difference in depth values is calculated between adjacent points. If the difference is less than the threshold, the two points are on the same surface of the human body. Otherwise, they are in different parts of the human body.

3.2.2. Geometric Feature

Geometric features refer to the overall attributes, common ones include texture and shape features of the human body. To eliminate the influence of complex poses by constructing and merging 3D point clouds of multiple views. The part detector was used to detect the body parts [51], and then the centroid of each part was obtained as the joint position. Based on the shape segmentation and skeleton sequence, Zhang et al. [52] designed an extraction method of human skeleton. In the preliminary step, the centroid of every part was also used to generate a pseudo skeleton. Multiple depth sensors were also utilized to achieve the purpose of motion capture [53]. First, multi-frame depth data from the depth sensor was converted into multiple point clouds, and then, these point clouds were combined into a merged point cloud, on which the skeleton line was acquired by the Reeb graph. Finally, the joint position was calculated from the skeleton line according to the joint structure of the human body. A curve skeleton expression based on the set of cross-section centroids was presented [54]. Patil et al. [55] applied multiple inertial measurement unit (IMU) sensors, which were placed at the human joints to estimate the 3D position of the joints, the Lidar data compensate for displacement drift during the initial calibration of the skeleton structure. A 2.5D thinning algorithm was exerted [56], including segmentation of the occlusion region and thinning line extraction. The thinning line bone obtained cannot determine the exact position of all body joints, but the end-joints of the body part can be detected. Finally, it was registered with the constructed human model containing 16 bone joints, and the human joints were extracted.

Xu et al. [57] detected the human joints in a single-frame point cloud using the TOF depth camera. The process was distributed into three stages as shown in Figure 9. An in-house captured 3D dataset containing 1200-frame depth images was first collected, which can be categorized into four different poses (upright, raising hands, parallel arms, and akimbo). To eliminate the influence of the background and noise points on the algorithm, the point cloud was separated from the background by the conditional filtering in the data pre-processing stage. To avoid self-occlusion, the point cloud was projected to the 2D top view, and then the point cloud was easily rotated by the angle, which was formed by the farthest points on the x-axis and the horizontal axis, to make the viewpoint of the camera parallel with the direction of the human body being facing. Finally, the 3D silhouette of the human body was extracted by adopting the public algorithm in the PCL [7] as a global feature in the point cloud.

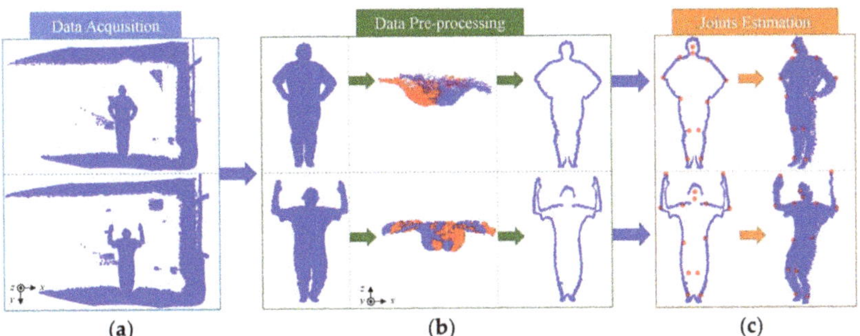

Figure 9. Geometric feature is used for human joint estimation. The approach consists of three stages: data acquisition, data pre-processing and joint estimation. (**a**) The point clouds are directly obtained from the depth camera. (**b**) Data pre-proposing mainly involves three parts: firstly, the irrelevant points are filtered, then the orientation of the human point cloud is adjusted, and finally the 3D silhouette is extracted. (**c**) Fourteen joints of human body are extracted by using the geometric feature of human silhouette. Figure from [57].

Before extracting the joints of human body, different poses were classified according to the angle and aspect ratio of the silhouette point cloud. First, the four poses are divided into two categories, according to the angle formed by the farthest point on the x-axis and the point with minimum value on the y-axis, one includes upright and akimbo poses, and the other contains the remaining poses. Then, the two poses in each category are further distinguished according to the aspect ratio of the silhouette. There was slight difference in the extraction method of the 14 joints for different poses. The approximate flow was that the head and foot joints were regarded as the centroids of each segmented part according to body proportion. As the base point, the waist joint was obtained in the next step using the prior information, while the shoulder and hand joints can be acquired afterwards. The elbow and knee joints were calculated by judging whether bending was present. In an upright state, the elbow joint was determined to be the midpoint between the hand joint and the shoulder joint, while the knee joint was also located on the line between the foot point and the midpoint of the left and right shoulders. When in the bent state, the elbow joint was defined the farthest point in the arm point cloud from the straight line formed the hand and shoulder joints, and the knee joint was the minimum value in the z direction.

Compared with the other methods, the accuracy of the joints was greatly improved. The average joint error was less than 5.8 cm by using both the in-house and public datasets, but it was also affected by the clothes, which led to more error in the waist joint.

3.3. Machine Learning-Based Methods

Given the rapid development of machine learning technology in computer vision, some of the latest deep learning networks, such as PointNet [58], VoxelNet [59], PointCNN [60] and PointConv [61], are also implemented in the 3D point clouds. These algorithms have further pushed the development of deep learning on 3D point clouds to address various problems [62,63]. This review attempts to track and summarize the progress of point cloud-based networks for human tracking in recent years, so as to provide a clear prospect for the current point cloud-based joint extraction of the human body. We mainly summarize from two categories of neural network and classification tree.

3.3.1. Neural Network

One very important field of machine learning is the neural networks. Especially, convolutional neural network (CNN) is a fascinating and powerful tool that can achieve great analysis results in many tasks of computer vision. A 2D CNN is used to locate 2D human joints, which are then extended to 3D through a depth transformation to reduce the computational cost. Biswas et al. [64] designed an end-to-end system that combines

RGB images and point cloud information to recover 3D human pose. Özbay et al. [65] used a simplified extraction method "Conditional Random Field" to classify 3D human point clouds, and the corresponding images and poses as input of CNN transmitted similar spaces. When the image-pose pair is matched, the value of dot product is high, otherwise the value is low. Without making any assumption about the appearance and initial pose of the human, the proposed system could be applied to multi-human interaction scenarios [66]. Schnürer et al. [67] utilized networks to generate 2D belief maps, combined with depth information for pose detection of the upper body, which required fewer resources while achieving a high frame rate. However, the depth mapping of a 2D single-channel image did not represent an actual 3D representation. To overcome this limitation, a 3D CNN architecture was proposed to provide a likelihood map for each joint [68], and the detection structure was extended to make it suitable for multi-person pose estimation. Millimeter wave (mmWave) has the advantages of high bandwidth and fast speed, which is the reason why it is used as the carrier of 5G technology. A new method of real-time detection and tracking of human joints using mmWave radar was proposed [69], named mmPose. This is the first method to detect different joints using mmWave radar reflected signals, and the emission wave at 77 GHz allowed it to capture small differences from the reflective surface. The algorithm structure is shown in Figure 10.

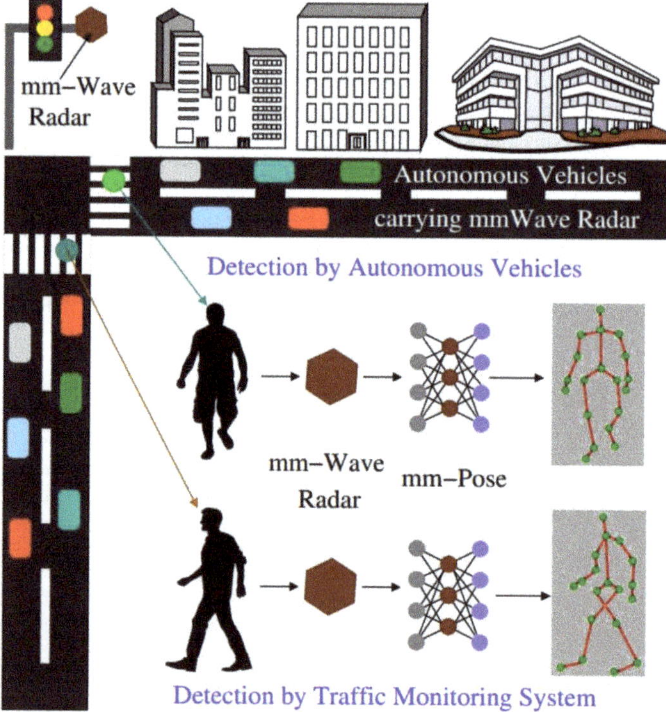

Figure 10. In automatic/semi-autonomous vehicles and traffic monitoring systems, mm-Pose can be used to perform robust skeleton pose estimation of pedestrians. Figure from [69].

The objects reflected the radar signal within a coherent processing interval (CPI), and a 3D radar cube was obtained with fast-time, slow-time and channel. In order to overcome the sparseness of the voxel grids. and significantly reduce the subsequent machine learning structure, the depth, the ratio between elevation and azimuth, and normalized power values of the reflected signal were assigned to the RGB channels to generate a 3D heat map, which can be used as the input to CNN, and the output of CNN were different human joints in 3D space.

In addition to CNN, other neural networks are also commonly used in point cloud-based pose recognition. Fully connected network (FCN) was introduced to accurately simulate the restriction of the human joint [70], which can effectively implement the realistic restriction by transforming the constraint force in the physics engine into an optimization problem. Li et al. [71] proposed multi-layer residual network to obtain hand features for tracking and segmenting. Zhang et al. [63] adopted an adversarial learning method to ensure the effectiveness of the restored human pose to alleviate the ambiguity of the human pose caused by weak supervision. A deep learning-based weakly supervised network, as shown in Figure 11, not only used the weakly-supervised annotations of 2D joints, but also applied the fully supervised annotations of 3D joints. It is worth noting that 2D joints of human body can help select effective sampling points to reduce the computational cost of the point cloud-based network.

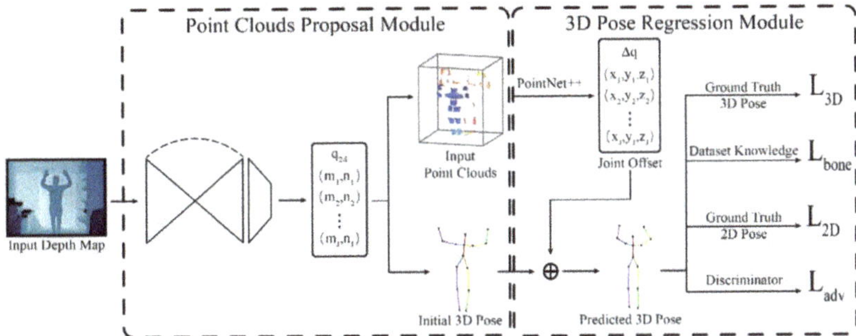

Figure 11. A schematic diagram of human joint estimation using Neural Network. The network consists of two modules, the point clouds proposal module and the 3D pose regression module. Using the input depth map, we first estimate the 2D human pose, and use it to sample and normalize the extracted point clouds from depth. Then we use the initial 3D pose converted from the estimated 2D pose and the normalized point clouds to predict the final 3D human pose. Figure from [63].

In this paper, a point cloud-based network is involved. Initially, Qi et al. [58] proposed the PointNet network, which can extract features of point from unordered point clouds. PointNet used traditional multilayer perceptrons (MLPs) as the core learning layer. It is commonly used to deal with 3D object classification and point-level semantic tasks. In the subsequent research work, PointNet++ [72] added local structures at different scales to enhance PointNet. Because of the effectiveness of this method, the author used the PointNet++ network to deal with point segmentation. Compared with the existing methods about pose estimation of the human body that require human foreground detection, this method can perform accurate pose estimation without clear requirements.

Joint extraction of part structures from human body attracted much attention for further research. The proposed self-organizing network aims to use unannotated data to obtain accurate 3D hand pose estimation [73]. The heat map, as the output of 3D CNN, reflected the probability distribution of the joints. In [74], heat map was used as the intermediate supervision of the 3D hourglass network to participate in the skeletal constraints for the hand tracking. In addition to the heat map representing the distance, the unit vector field was introduced, and joint position was inferred by weighted fusion [75]. In order to further improve the accuracy of the fingertips, a fingertip refinement network was designed to model the visible surface of the hand and perform pose regression [76]. Different from the original PointNet, Local Continuous PointNet (LCPN) [77] was proposed to extract the local features of the neighbor index in the unorganized point cloud to estimate the facial joints. The input of the 3D CNN was encoded through the projection of the point cloud [78]. After the convolution and pooling layer, 3D features can be extracted from the volume representation, which can be used to return the relative position of the hand joints in the 3D volume. An end-to-end multi-person 3D network Point R-CNN for the pose estimation was proposed [79], which used panoramic point clouds of multiple cameras to solve the occlusion problem. The whole network can be regarded as a combination of two parts. The first segment is for the instance detection using VoxelNet and the other segment is for instance processing by the PointNet to acquire the joint information.

3.3.2. Classification Tree

The classification tree is one of the prevailing methods for human body segmentation. As a newly emerging and highly flexible machine learning algorithm, random forest (RF) refers to a classifier that uses multiple trees to train and predict samples, has a wide range of applications.

Inspired by the decision forest, each point in the point cloud of the human body is voted to evaluate the contribution of each part of the human body, so that collaborative method was proposed to learn the 3D features of the human body [80]. Xia et al. [81] trained the cascade regression network from the pre-recorded human motion dataset. In addition, the hierarchical kinematics model of the human pose was introduced into the learning process, it can directly estimate the accurate angles of the 3D joints.

Random forest is often used to segment different parts of the human body in the following literatures. Different regions of the upper body were first detected, and then the probability map for each region were calculated [82]. The highest part in the probability map was defined as the external joints. The internal joints, such as the elbow, were fitted with an ellipse model to obtain. In the 3D point cloud, the Principal Direction Analysis (PDA) was used to estimate the main direction of the body part, and then the main direction was mapped to each part of the 3D model to estimate the human pose [83]. In the prescribed action set, a pose estimation using multiple random forests was proposed to enhance the results of motion analysis [84]. A group of random verification forests were set to verify classification results of the initial random regression forest for precise joints positioning. The geodesic-based feature descriptors played a significant role in the random forest classifier to produce more exact spatial predictions for body parts and bone joints [85]. Random forest was also applied to infer the consistency between the input data and the construction template [86]. The method successfully restores the shape of the human body and extracts joints.

The method of pixel inference using random decision tree usually requires more heavy calculation. Especially when the number of trees is increased to improve generalization and accuracy, the computational burden of multiple trees may force a trade-off between speed and accuracy, and the random tree walk (RTW) method can obtain greater gain. The method combined RTW with optimization methods such as ICP and random search, which raised the ability to extend of the classification tree [87]. RTW was used to initialize various assumptions in different ways and then passed them to the optimization stage.

Yub et al. [88] no longer trained the tree for pixel-level classification, and used the regression tree to estimate the probability distribution towards a specific joint direction relative to the current position. In the test process, the direction of random walking was randomly selected from a set of representative directions. A new position by a constant step was found in that direction.

For all positioning problems, as long as we know the direction of any point on the object towards that position, we can find the correct position. Ideally, the orientation of all parts should be trained from all possible positions of the whole body, because random tree walking could reach the joint position faster, so a starting point close to the target joint position was required. In the case of using the skeleton topology, one needed to provide a nearby initial point for the RTW, as shown in Figure 12.

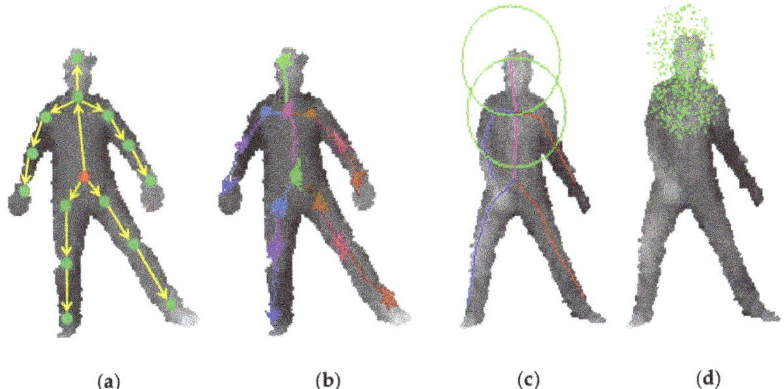

(a)   (b)   (c)   (d)

Figure 12. Example of classification tree for human joint estimation. (**a**) illustrates the kinematic tree implemented along with random tree walk (RTW). First, the random walk toward belly positions starts from body center. The belly positions (red dot in (**a**)) become starting point for hips and chest, and so forth. (**b**) shows the RTW path examples. (**c**) illustrates offset sample range spheres in green. In (**d**), the green dots represent offset samples. Figure from [88].

RTW can be described as training regression trees for each joint in the human skeleton. Here, the direction from the point to that specific joint is obtained by training a regression tree. Therefore, a training set is first constructed with the position of each joint point and the depth value of the input point.

The unit direction vector $\hat{u}$ from the offset point to the joint was defined as Equation (13):

$$\hat{u} = \left(p_j - q\right) / \|p_j - q\|, \tag{13}$$

where p_j is the coordinate of a random point, q is the position of the specific joint. The training sample S is expressed as Equation (14):

$$S = (I, q, \hat{u}), \tag{14}$$

I represents the depth value. The goal is to find a partitioned binary tree that minimizes the sum of squared differences. At the same time, the directions are stored on each leaf node in the form of clusters, so that several representative directions and corresponding probability weights form the output of the tree. When estimating the pose, the path starts walking randomly from some initial points. In each step of the traversal, the regression tree is traversed to a leaf node, where a set of directions corresponding to the current point can be obtained. However, the step direction is randomly selected from the k-means cluster unit vector at the leaf node.

3.4. Summary

The template-based method first needs to establish a template library or a parameterized template, and then the similarity between the point cloud of human body and the sample in the template library or the target model is compared. This method is relatively rough and time-consuming. Given diversity and multi-scale structure of the sample data, the same pose of human body may be very different in space. Therefore, the accuracy of template-based methods is very limited.

Feature-based method needs to extract the global or local features of the point cloud, which combine with some prior knowledge to obtain the 3D joints of the human body. This method relies on the selection of feature points such that it is not suitable for self-occlusion and changing poses. Therefore, it is necessary to further optimize the robustness of the algorithm for covering the poses of the human body as much as possible.

The machine learning-based method mainly uses the network to automatically learn the required features from the point cloud, and then the learned features can be regarded as the judgment condition to extract the human joints. Compared with the above two methods, it has been greatly improved. On one hand, the obtained joints can achieve higher accuracy by learning sample features in a large training set, and on the other hand, it is also very robust to scale processing. The machine learning-based method can make up for the shortcomings of the above two methods, but it is restricted by the sample richness of the training set, so the construction of the training set is very important to the machine learning-based method.

Moreover, we summarize some works of point cloud-based joint estimation for human body in Table 2. For 3D pose estimation, two different error metrics can estimate the accuracy of the method. One is direct measurement of the Euclidean distance between the estimated and ground truth joints, another is the average precision (AP), which is defined as the ratio of correctly estimated joints within a specific threshold. Some works have adopted the AP error metrics, and Table 2 reports datasets together with some key parameters, especially, the threshold values δ. When δ changes, the AP of each joint would be different. The 3D coordinate of each classified joints, which are obtained by proposing algorithm in the literature, is compared with the corresponding ground truths of the joints within the same dataset. When the difference between the two values above is less than the given threshold δ in Table 2, the joint coordinate is considered to be a correct position. For acquiring the tracking accuracy of each classified joint in the referenced works, the ratio AP in Table 3 is finally calculated between the sum of the correct locating joints and all joints.

Table 2. Summary of the referenced works for human pose estimation with depth inputs.

Methods	Dataset	δ (/cm)	FPS	GPU
[11]	SMMC-10	9	25	N
[18]	EVAL	10	30	Y
[24]	SMMC-10	10	20	N
[89]	SMMC-10	10	125	N
[44]	SMMC-10	6	25	N
[45]	Self-built dataset	-	-	-
[50]	Self-built dataset	6	-	-
[56]	SMMC-10	15	25	N
[57]	Self-built dataset	6	-	N
[68]	EVAL	10	-	Y
[82]	CMU Mocap	-	-	Y
[87]	Self-built dataset	10	35	N

"-" represents that the value is not given; δ represents threshold value; "Y" means running on a desktop with GPU; and "N" means running on a desktop without GPU.

Table 3. Tracking accuracy of the human body joints in the referenced works.

	Head	Neck	Shoulders	Elbows	Wrists	Hand	Ankles	Knees	Foot
[11]	0.97	-	-	-	-	0.85	-	-	-
[18]	0.91	-	0.94	0.905	0.818	-	0.93	0.955	-
[24]	0.99	-	0.965	0.965	0.965	-	0.97	0.958	-
[89]	0.975	0.965	0.985	0.96	0.95	-	0.965	0.963	-
[44]	0.96	-	0.995	0.774	-	0.933	-	-	0.99
[45]	0.92	-	-	-	-	0.852	-	-	0.869
[50]	-	0.813	0.882	0.867	-	-	-	0.85	
[56]	0.97	-	0.935	0.86	-	0.885	-	0.935	0.935
[57]	0.979	0.701	0.947	0.926	-	0.754	-	0.463	0.855
[68]	0.917	0.955	0.919	0.763	-	0.839	-	0.852	0.927
[82]	0.97	-	0.955	0.915	0.867	-	0.930	0.925	-
[87]	0.99	-	0.975	0.96	0.95	-	0.965	0.98	-

"-" represents that the value is not given.

4. Public Depth Dataset of the Human Body

Public datasets play an important role in testing the robustness of the algorithm and provide a platform to compare different algorithms in a fair manner. In the past few years, many 3D benchmark datasets for different applications have been collected and made publicly available to the research community. The structure of the dataset mainly includes RGB, depth map or point cloud acquired from structured light or ToF depth camera. In this paper, we only focus on the point cloud-based dataset. This section provides a detailed review of the datasets listed in Table 4. Since point cloud occupies fairly large storage space, most datasets usually provide depth maps together with the internal parameters of the camera, which can be easily converted to the point cloud-based datasets.

Table 4. Previously developed depth datasets for human bodies.

Dataset Name	Acquisition Device	Year	Joints	Subjects	Classes	FPS
SMMC-10 [90]	MoCap+ToF	2010	15	S	-	25
MSR-Action3D [91]	Kinect v1	2010	20	S	20	15
NTU RGB+D [92]	Kinect v2	2016	25	M	60	30
Walking gait [93]	Kinect v2	2018	25	S	9	30
MHAD [94]	MoCap+Kinect v1	2013	21	S	11	30
G3D [95]	Kinect v1	2012	20	S	20	30
G3Di [96]	Kinect v1	2014	20	M	18	30
SBU-Kinect-Interaction [97]	Kinect v1	2012	15	M	8	15
CDC4CNV [98]	Kinect v1	2011	9	S	-	-
EVAL [89]	Vicon motion	2012	12	S	24	30
CMU MoCap [99]	MoCap	-	41	M	23	-

S denotes single person, M denotes multi-person interaction. "-" represents that the value is not given.

A set of widely availed depth dataset, named SMMC-10, was constructed as a benchmark for the algorithm testing [90]. To generate this dataset, a probability model containing 15 rigid parts of the human body was first defined. These rigid parts were spatially constrained by the joints with 48 degrees of freedom. This dataset was recorded by the Motion Capture (MoCap) system and the ToF camera (Swissranger SR4000) at 100–250 ms per frame. It included 28 real actions, such as fast kicking, swinging, self-focusing, and whole-body rotation.

Another constructed dataset by Li et al. [91], named MSR-Action3D (https://documents.uow.edu.au/~wanqing/#MSRAction3DDatasets (accessed on 12 June 2020)), was 20 game action for seven subjects facing the depth camera, including: high arm wave, horizontal arm wave, hammer, hand catch, forward punch, high throw, draw x, draw tick, draw circle, hand clap, two hand wave, side-boxing, bend, forward kick, side kick, jogging, tennis swing, tennis serve, golf swing, and pickup and throw. Each action was captured three times by Kinect v1 (Microsoft Corp., Redmond, WA, USA) at 15 frames per second. In total, the dataset reasonably covered various movements of arms, legs, and torso, which stored 4020 motion samples with 23,797 depth maps. Notes that, if an action was done with only one arm or one leg, subjects were advised to use their right arms or right legs.

A large-scale RGB+D human action recognition dataset, named NTU RGB+D dataset (http://rose1.ntu.edu.sg/Datasets/actionRecognition.asp (accessed on 12 December 2020)), used a recurrent neural network (RNN) to simulate the long-term time correlation of various parts in the human body, and to better classify the human body poses [92]. The dataset collected more than 56,000 video samples with a total of four million frames from 40 different subjects and 60 different operation classes, including daily operations, interoperations, and health-related operations.

Nguyen et al. [93] explored the extraction of skeleton during human walking. The content of the walking gait dataset was about 18.6 GB, and it was divided into a test set and a training set. The test set included samples of five subjects, and the training set was walking gait data of four individuals. The data collected for each person includes the information of skeleton and silhouette together with the point cloud.

Berkeley Multimodal Human Action Database (MHAD) took seven males and five females aged between 23 to 30 years old as subjects (http://tele-immersion.citris-uc.org/berkeley_mhad (accessed on 2 November 2020)) [94]. Each subject performed 11 actions in succession, such as jumping, throwing, waving, sitting down and so on. To ensure the accuracy of action acquisition, all subjects repeated 5 times for each action, resulting in a total of approximately 660 action sequences. In addition, a T-pose model was created for each subject to extract its corresponding skeleton.

The G3D dataset (http://dipersec.king.ac.uk/G3D/G3D.html (accessed on 29 June 2020)) mainly includes different game actions [95]. Given the internal parameters of the depth camera, the captured depth map can be converted into a point cloud. The dataset contains 10 subjects. Each subject was required to complete 7 action sequences consisting of 20 game actions: punch right, punch left, kick right, kick left, defend, golf swing, tennis swing forehand, tennis swing backhand, tennis serve, throw bowling ball, aim and fire gun, walk, run, jump, climb, crouch, steer a car, wave, flap and clap. On this basis, G3Di (http://dipersec.king.ac.uk/G3D/G3Di.html (accessed on 29 June 2020)) [96] was constructed to process a human interaction for multiplayer games. The dataset contained six pairs of subjects' motion interaction behaviors, such as boxing, volleyball, football, table tennis, sprinting and hurdles. Each action was separately stored as RGB, depth and skeleton data.

A complex human activity dataset, called SBU-Kinect-Interaction (https://www3.cs.stonybrook.edu/~kyun/research/kinect_interaction/index.html (accessed on 19 December 2020)), was created to describe the interaction between two people [97], including synchronized video, depth and motion capture data. All videos were recorded in the same laboratory environment. Seven participants performed the activity consisting of 21 groups, where each group contained a pair of different people performing all eight interactions. Note that in most interactions, one person was acting and the other was reacting. Each action category contained one or two sequences. There were approximately 300 interactions in the entire dataset.

The CDC4CV pose dataset [98] was acquired with the depth information of the upper body for comparison of static pose estimation techniques by the Kinect v1, including 9 joints of three subjects. During the acquirement of the depth pose of human body, the upper body of each subject was ensured to stay within the 640×480 window. Nearly 700 depth data including three subjects were collected and labelled, of which 345 depth data were chosen as the training set and the rest data were used as the test set.

The EVAL dataset was built in 2012 (http://ai.stanford.edu/~varung/eccv12 (accessed on 16 September 2020)) [89], which included 24 action sequences of three different subjects. Each subject performed actions of gradually increasing complexity at the place where is approximately 3 m away from a Kinect camera. The ground truth of the 12 joints was captured using the Vicon motion capture system, and stored in the EVAL dataset together with the corresponding 3D point clouds.

CMU MoCap (http://mocap.cs.cmu.edu (accessed on 14 October 2020)) [99] used 12 Vicon infrared MX-40 cameras to collect the motions of the human body wearing black jumpsuits, including six major categories, such as human interaction, interaction with environment, locomotion, physical activities and sports, situations and scenarios, and test motions. And each category was further divided into 23 sub-categories. 41 marks were posted on the human body for the cameras to collect the ground truth of the joints during the motion. The images captured by various cameras were then triangulated to obtain 3D data.

In summary, MoCap is a motion capture system by posting marks on the joints of the human body with multiple cameras to track the human joints from different views. Accurate 3D skeleton information at a very high frame rate is acquired in the system. However, the system is usually expensive, and only available in an indoor environment. At present, many methods also use a single depth camera for data acquiring and processing. The subjects do not need to wear any equipment with constraints. For the single person datasets, MSR-Action3D and G3D used gaming action as the main application. Both of them were single view and in the similar action sequence. In addition to depth data, MSR-Action3D also collected the video data, and G3D provided the corresponding RGB images at the relatively high frame rate. SMMC-10, Walking gait, MHAD, CDC4CNV and EVAL are mainly included basic behaviors of the human body. Only single individual was used to complete a series of complex actions in SMMC-10. MHAD contained the depth information of 12 subjects from four different views; the gender and age of the subjects were also given. EVAL provided the ground truth of 12 joints and the corresponding information of the 3D point cloud. CDC4CNV tracked the nine joints of the upper body, and Walking gait was used to analyze human gait, both of their application scenarios have certain limitations. For the multi-person datasets, the MoCap system can be used to obtain the ground truths of 41 joints in CMU MoCap dataset, it covered multi-person gaming, sport and other behaviors. SBU-Kinect-Interaction provided eight classes of interaction sequences. After that, G3Di provided common interaction activities for multi-person gaming, 20 joints of the human body were given for detailed analysis. NTU RGB+D used Kinect v2 to acquire the ground truths of 25 joints with more human interaction activities. In addition to the depth information, the dataset also included RGB images and IR videos.

5. Application of Point Cloud-Based Joint Estimation

Human joint recognition is one of the important directions of artificial intelligence applications. With the maturity of technology, human-related research can use joint information to solve some problems. According to different application scenarios, the approaches can be divided into the following categories, including virtual try-on technology, 3D human reconstruction, action recognition, human–computer interaction, and many others, some examples of the above application are shown in Figure 13. The related literatures are summarized according to the above application scenarios.

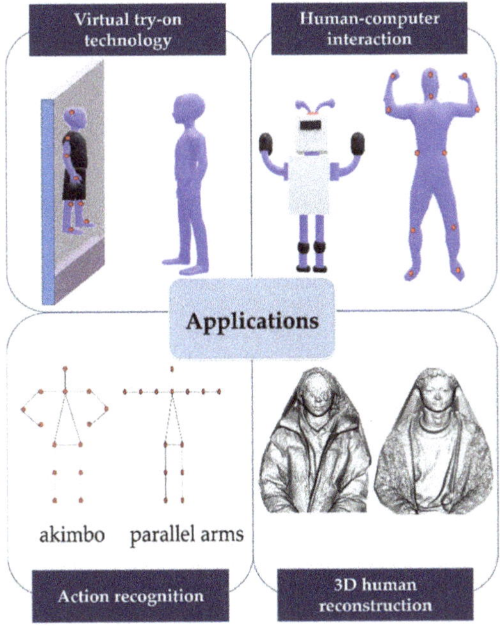

Figure 13. Applications of point cloud-based joint estimation.

Human body shape estimation is essential for virtual try-on technology. Estimating the 3D human shape in motion from a set of unstructured 3D point clouds is a very challenging task. Human joints can play an important role as a priori in 3D shape estimation of human body. Yang et al. [100] proposed an automatic method to solve the estimation of the human body shape in motion. Under the premise of wearing loose clothes, the model reconstruction problem is expressed as an optimization problem by controlling the body shape. Based on the automatic detection of human joints, the pose fitting scheme was optimized [101]. The results of 3D scanning from multiple viewpoints were projected onto 2D images, and then deep learning algorithms were utilized to mark the joints, which were helpful to find the best pose parameters. With the help of the joints in the SMPL model and manually marked the joints in the point cloud for registration, the result of coarse registration was obtained, and the hot core feature was extracted between the two frames during the changing pose for non-rigid registration. Both the result of non-rigid registration and coarse registration was fitted each other to get the final 3D human body model [102]. The joints of the frontal point cloud of the human body, which was generated directly with Kinect device, helped initialize the personalized SMPL model, and the model was registered with the input point cloud to find the corresponding points for obtaining a 3D human body model [103]. Yao et al. [104] further projected the obtained model onto the corresponding RGB image.

The joints can also assist the 3D reconstruction of the human body. Matteo et al. [1] utilized the information provided by the skeletal tracking algorithm to transform each point cloud into a standard pose in real time, and then registered each transformed point cloud to achieve 3D human body reconstruction. In order to extract more point features, a graph aggregation module was used to enhance PointNet++ [2], an attention module was used to better map disordered point features into ordered skeleton joint, and a skeleton graph module aimed to regress the skeleton joints by SMPL parameters. A dataset containing 2D scenes and 3D human body models was constructed. After marking the joints of human body on 2D image, they were converted to 3D coordinate system generated by the radar, and the SMPL model was used to fit the pose of the human body [105].

To improve the accuracy and real-time performance of action recognition, the skeleton-based method is studied in various research fields as an effective technology. Instead of using the entire skeleton as the input to Hierarchical RNN, the human skeleton was divided into five parts according to the physical structure of the human body, and then they were fed into five subnets, respectively. As the number of layers increases, the representation extracted by the subnet merges as higher layer output [106]. Through the distance of the joints and occupancy information of the skeleton, the time information was also extracted using the time pyramid to form the dataset of each action [107,108]. To recognize human actions, the difference in values between consecutive frames was used to calculate the new positions and angles of all joints. The input of the structure tree neural network was the human joints, and the output was the action classification [109]. Zhang et al. [110] extracted the local surface geometric feature (LSGF) of each joint in the point cloud, and introduced the global feature of the vector encoding video sequence. Finally, the SVM classifier was applied to reach the result of the action classification. Khokhova et al. [111] utilized a regular grid to divide the 3D space, and then used descriptors based on space occupancy information to identify the pose of the static frame. An enhanced skeleton visualization method was present [112], in which a CNN was implemented as the main structure to recognize the view-invariant human action.

Human–robot motion retargeting is one of interesting research in human–computer interaction technology. The goal of human–robot motion retargeting is to make the robot follow the movement of the human body. Wang et al. [113,114] established a model as a bridge between the input point cloud of the human body and the robot, so as to achieve the purpose of human–robot motion retargeting. The activity was decomposed into multiple unit sequences, each unit was related to an important factor of behavior [115], and then was inputted into a dynamic Bayesian network to analyze human behavior intentions and realize human–computer interaction.

In addition to some of the above applications, Kim et al. [116] used high-speed RGB and depth sensors to generate movement data of an expert dancer; all skeletons could be reorganized to generate desired dance movements. Given the visual input, let the robot ratiocinate and choose the best container and human pose to perform a transfer task [117]. Soft biometrics can solve the problem of people re-identification. For each measured subject, the 3D skeleton information was applied to adjust the human pose and create the standard pose (SSP) of the skeleton. The SSP was divided into grids to obtain individual characteristics for identification [118]. In archaeology, skeleton joints are helpful to generate a model that can represent any biological shape [119]. The length and position of joints were also beneficial to judge whether the age of the human body is a child or an adult [120]. Desai et al. [121] used the direction of the foot or torso to judge the orientation of the human, and then combined and optimized the skeletons collected by multiple cameras to obtain the final skeleton, even in the case of occlusion. In terms of rehabilitation treatment, a method was advanced to improve the evaluation of upper-limb rehabilitation [122]. The skeleton of the point cloud was taken by Microsoft Kinect, which was registered with the SMPL template to obtain the position and length of the joint.

6. Conclusions

In this paper, we review recent researches on the point cloud-based joint extraction of human body. The superiority of point cloud data as well as the applications of joint estimation are all discussed in details. Different works are introduced based on three mainstream methods: (1) template-based methods; (2) feature-based methods; (3) machine learning-based methods. On this basis, we analyze and summarize the current human pose dataset with point cloud. Although a lot of research devotes to the construct the practical pose dataset of human body, there is still a lack of comprehensive ground truth datasets for human varied pose, especially the marking of human joints with different clothes. The relevant applications of point cloud-based joint estimation of human body are discussed in this paper, we found that point cloud-based method plays an important role in some emerging technologies, such as 3D reconstruction, human–computer interaction, action recognition, etc.

From the analysis above, we know that many existing methods have already accurately tracked the human body joints in real time under the indoor environment. However, joint estimation of human body yet still faces many challenges. In our opinion, feature-based method cannot further improve the accuracy of joint detection, if it relies only on the depth features and length constraints of the joints. Therefore, combining with other data from multiple sensors has become a new breakthrough, such as RGB cameras, infrared cameras and IMU sensors. The template-based method and the machine learning-based method are currently unable to recognize the joints of any pose, because neither the template dataset nor the training set can cover all the poses. The 3D template of the human body can be constructed by some software, but there are some limitations on the fixed pose of the template, and the matching process takes a long time. At present, additional information can be leveraged to shorten the search time, it is also possible to build models with better resolution. They will still be interesting research directions in the coming future. In order to improve the detection accuracy, the machine learning-based network should output the local and global features between the points in real time when the input is an orderless point cloud. In addition, there are still some unresolved challenges and gaps between research and practical applications in the entire research field, such as self-occlusion and multi-person detection. However, with the deeper research of machine learning technology, pose estimation of the human body will also be faster and more accurate. Effective networks and sufficient train data are key elements in machine learning-based methods; it is believed that there will be more room for improvement by many scientific researchers investing time in the future.

Author Contributions: Conceptualization, Y.Y.; methodology, T.X., D.A., and Y.Y.; validation, T.X., D.A., and Y.J.; formal analysis, T.X., D.A., and Y.Y.; investigation, T.X., D.A., and Y.J.; data curation, T.X., D.A., and Y.J.; writing—original draft preparation, T.X., D.A.; writing—review and editing, Y.Y.; visualization, D.A., T.X.; supervision, Y.Y.; project administration, Y.Y.; funding acquisition, Y.Y. All authors have read and agreed to the published version of the manuscript.

Funding: This work was supported in part by the National Key Research and Development Program of China under Grant 2018YFB0703500, in part by the Key Technologies Research and Development Program of Tianjin under Grant 20YFZCGX00440, Tianjin Research Innovation Project for Postgraduate Students (Artificial Intelligence Special Project) under Grant 2020YJSZXB09, and in part by the Fundamental Research Funds for the Central Universities, Nankai University, under Grant 63201178 and Grant 63191511.

Institutional Review Board Statement: Not applicable.

Informed Consent Statement: Not applicable.

Data Availability Statement: No new data were created or analyzed in this study. Data sharing is not applicable to this article.

Acknowledgments: Valuable feedback and suggestions for improvements of this paper from Qiang Wang and Zhongqi Pan are gratefully acknowledged.

Conflicts of Interest: The authors declare no conflict of interest.

References

1. Munaro, M.; Basso, A.; Fossati, A.; Van Gool, L.; Menegatti, E. 3D reconstruction of freely moving persons for re-identification with a depth sensor. In Proceedings of the 2014 IEEE International Conference on Robotics and Automation (ICRA), Hong Kong, China, 31 May–7 June 2014; pp. 4512–4519.
2. Jiang, H.; Cai, J.; Zheng, J. Skeleton-aware 3d human shape reconstruction from point clouds. In Proceedings of the IEEE/CVF International Conference on Computer Vision, Seoul, Korea, 27 October–2 November 2019; pp. 5431–5441.
3. Jalal, A.; Kamal, S.; Azurdia-Meza, C.A. Depth maps-based human segmentation and action recognition using full-body plus body color cues via recognizer engine. *J. Electr. Eng. Technol.* **2019**, *14*, 455–461. [CrossRef]
4. Park, S.; Park, J.; Al-Masni, M.; Al-Antari, M.; Uddin, M.Z.; Kim, T.-S. A depth camera-based human activity recognition via deep learning recurrent neural network for health and social care services. *Procedia Comput. Sci.* **2016**, *100*, 78–84. [CrossRef]
5. Xu, H.; Yu, Y.; Zhou, Y.; Li, Y.; Du, S. Measuring accurate body parameters of dressed humans with large-scale motion using a Kinect sensor. *Sensors* **2013**, *13*, 11362–11384. [CrossRef] [PubMed]
6. Meng, Y.; Jin, Y.; Yin, J. Modeling activity-dependent plasticity in BCM spiking neural networks with application to human behavior recognition. *IEEE Trans. Neural Netw.* **2011**, *22*, 1952–1966. [CrossRef]
7. Rusu, R.B.; Cousins, S. 3d is here: Point cloud library (pcl). In Proceedings of the 2011 IEEE international conference on robotics and automation, Shanghai, China, 9–13 May 2011; pp. 1–4.
8. Zhou, Q.-Y.; Park, J.; Koltun, V. Open3D: A modern library for 3D data processing. *arXiv* **2018**, arXiv:1801.09847.
9. Knoop, S.; Vacek, S.; Dillmann, R. Sensor fusion for 3D human body tracking with an articulated 3D body model. In Proceedings of the 2006 IEEE International Conference on Robotics and Automation (ICRA), Orlando, FL, USA, 15–19 May 2006; pp. 1686–1691.
10. Zhu, Y.; Dariush, B.; Fujimura, K. Kinematic self retargeting: A framework for human pose estimation. *Comput. Vis. Image Underst.* **2010**, *114*, 1362–1375. [CrossRef]
11. Suau, X.; Ruiz-Hidalgo, J.; Casas, J.R. Detecting end-effectors on 2.5 D data using geometric deformable models: Application to human pose estimation. *Comput. Vis. Image Underst.* **2013**, *117*, 281–288. [CrossRef]
12. Sethian, J.A. *Level Set Methods and Fast Marching Methods: Evolving Interfaces in Computational Geometry, Fluid Mechanics, Computer Vision, and Materials Science*; Cambridge University Press: Cambridge, UK, 1999.
13. Adalsteinsson, D.; Sethian, J.A. A fast level set method for propagating interfaces. *J. Comput. Phys.* **1995**, *118*, 269–277. [CrossRef]
14. Lehment, N.; Kaiser, M.; Rigoll, G. Using segmented 3D point clouds for accurate likelihood approximation in human pose tracking. *Int. J. Comput. Vis.* **2013**, *101*, 482–497. [CrossRef]
15. Sigalas, M.; Pateraki, M.; Oikonomidis, I.; Trahanias, P. Robust model-based 3d torso pose estimation in rgb-d sequences. In Proceedings of the IEEE International Conference on Computer Vision Workshops, Santiago, Chile, 7–13 December 2013; pp. 315–322.
16. Sigalas, M.; Pateraki, M.; Trahanias, P. Full-body pose tracking—the top view reprojection approach. *IEEE Trans. Pattern Anal. Mach. Intell.* **2015**, *38*, 1569–1582. [CrossRef] [PubMed]
17. Wu, Q.; Xu, G.; Li, M.; Chen, L.; Zhang, X.; Xie, J. Human pose estimation method based on single depth image. *IET Computer Vision* **2018**, *12*, 919–924. [CrossRef]
18. Ye, M.; Yang, R. Real-time simultaneous pose and shape estimation for articulated objects using a single depth camera. In Proceedings of the IEEE Conference on Computer Vision and Pattern Recognition, Columbus, OH, USA, 20–23 June 2014; pp. 2345–2352.

19. Xu, W.; Su, P.-c.; Cheung, S.-C.S. Human body reshaping and its application using multiple RGB-D sensors. *Signal Process. Image Commun.* **2019**, *79*, 71–81. [CrossRef]
20. Ge, S.; Fan, G. Non-rigid articulated point set registration for human pose estimation. In Proceedings of the 2015 IEEE Winter Conference on Applications of Computer Vision, Waikoloa, HI, USA, 5–9 January 2015; pp. 94–101.
21. Ge, S.; Fan, G. Sequential non-rigid point registration for 3D human pose tracking. In Proceedings of the 2015 IEEE International Conference on Image Processing (ICIP), Québec City, QC, Canada, 27–30 September 2015; pp. 1105–1109.
22. Ge, S.; Fan, G. Articulated Non-Rigid Point Set Registration for Human Pose Estimation from 3D Sensors. *Sensors* **2015**, *15*, 15218–15245. [CrossRef] [PubMed]
23. Stoll, C.; Hasler, N.; Gall, J.; Seidel, H.-P.; Theobalt, C. Fast articulated motion tracking using a sums of gaussians body model. In Proceedings of the 2011 International Conference on Computer Vision, Barcelona, Spain, 6–13 November 2011; pp. 951–958.
24. Ding, M.; Fan, G. Articulated gaussian kernel correlation for human pose estimation. In Proceedings of the IEEE Conference on Computer Vision and Pattern Recognition Workshops, Boston, MA, USA, 7–12 June 2015; pp. 57–64.
25. Ding, M.; Fan, G. Generalized sum of Gaussians for real-time human pose tracking from a single depth sensor. In Proceedings of the 2015 IEEE Winter Conference on Applications of Computer Vision, Waikoloa, HI, USA, 5–9 January 2015; pp. 47–54.
26. Ding, M.; Fan, G. Articulated and generalized gaussian kernel correlation for human pose estimation. *IEEE Trans. Image Process.* **2015**, *25*, 776–789. [CrossRef]
27. Oyama, M.; Aoyama, N.K.; Hayashi, M.; Sumi, K.; Yoshida, T. Two-stage model fitting approach for human body shape estimation from a single depth image. In Proceedings of the 2017 Fifteenth IAPR International Conference on Machine Vision Applications (MVA), Nagoya, Japan, 8–12 May 2017; pp. 234–237.
28. Zuffi, S.; Black, M.J. The stitched puppet: A graphical model of 3d human shape and pose. In Proceedings of the IEEE Conference on Computer Vision and Pattern Recognition, Boston, MA, USA, 8–10 June 2015; pp. 3537–3546.
29. Ye, M.; Wang, X.; Yang, R.; Ren, L.; Pollefeys, M. Accurate 3d pose estimation from a single depth image. In Proceedings of the 2011 International Conference on Computer Vision, Barcelona, Spain, 6–13 November 2011; pp. 731–738.
30. Grest, D.; Woetzel, J.; Koch, R. Nonlinear body pose estimation from depth images. In Proceedings of the 27th DAGM conference on Pattern Recognition, Vienna, Austria, 31 August–2 September 2005; pp. 285–292.
31. Park, B.-K.D.; Reed, M.P. A Model-based Approach to Rapid Estimation of Body Shape and Postures Using Low-Cost Depth Cameras. In Proceedings of the 8th International Conference and Exhibition on 3D Body Scanning and Processing Technologies, Montreal, QC, Canada, 11–12 October 2017; pp. 281–287.
32. Hesse, N.; Stachowiak, G.; Breuer, T.; Arens, M. Estimating body pose of infants in depth images using random ferns. In Proceedings of the IEEE International Conference on Computer Vision Workshops, Santiago, Chile, 7–13 December 2015; pp. 35–43.
33. Vasileiadis, M.; Malassiotis, S.; Giakoumis, D.; Bouganis, C.-S.; Tzovaras, D. Robust human pose tracking for realistic service robot applications. In Proceedings of the IEEE International Conference on Computer Vision Workshops, Venice, Italy, 22–29 October 2017; pp. 1363–1372.
34. Ye, G.; Liu, Y.; Deng, Y.; Hasler, N.; Ji, X.; Dai, Q.; Theobalt, C. Free-viewpoint video of human actors using multiple handheld kinects. *IEEE Trans. Cybern.* **2013**, *43*, 1370–1382. [PubMed]
35. Lu, X.; Deng, Z.; Luo, J.; Chen, W.; Yeung, S.-K.; He, Y. 3D articulated skeleton extraction using a single consumer-grade depth camera. *Comput. Vis. Image Underst.* **2019**, *188*, 102792. [CrossRef]
36. Huang, C.-H.; Boyer, E.; Ilic, S. Robust human body shape and pose tracking. In Proceedings of the International Conference on 3D Vision, Seattle, WA, USA, 29–30 June 2013; pp. 287–294.
37. Walsman, A.; Wan, W.; Schmidt, T.; Fox, D. Dynamic high resolution deformable articulated tracking. In Proceedings of the International Conference on 3D Vision, Qingdao, China, 10–12 October 2017; pp. 38–47.
38. Zhou, N.; Sastry, S.S. *Tracking of Deformable Human Avatars through Fusion of Low-Dimensional 2D and 3D Kinematic Models*; Technical Report UCB/EECS-2019-87; University of California: Berkeley, CA, USA, 19 May 2019.
39. Joo, H.; Simon, T.; Sheikh, Y. Total capture: A 3d deformation model for tracking faces, hands, and bodies. In Proceedings of the IEEE Conference on Computer Vision and Pattern Recognition, Salt Lake City, UT, USA, 18–22 June 2018; pp. 8320–8329.
40. Cao, C.; Weng, Y.; Zhou, S.; Tong, Y.; Zhou, K. Facewarehouse: A 3d facial expression database for visual computing. *IEEE Trans. Vis. Comput. Graph.* **2013**, *20*, 413–425.
41. Dijkstra, E. *A Discipline of Programming, volume 613924118*; Prentice-Hall Inc.: Englewood Cliffs, NJ, USA, 1976.
42. Krejov, P.; Bowden, R. Multi-touchless: Real-time fingertip detection and tracking using geodesic maxima. In Proceedings of the 2013 10th IEEE International Conference and Workshops on Automatic Face and Gesture Recognition (FG), Shanghai, China, 22–26 April 2013; pp. 1–7.
43. Phan, A.; Ferrie, F.P. Towards 3D human posture estimation using multiple kinects despite self-contacts. In Proceedings of the 14th IAPR International Conference on Machine Vision Applications (MVA), Tokyo, Japan, 18–22 May 2015; pp. 567–571.
44. Yuan, X.; Kong, L.; Feng, D.; Wei, Z. Automatic feature point detection and tracking of human actions in time-of-flight videos. *IEEE/CAA J. Autom. Sin.* **2017**, *4*, 677–685. [CrossRef]
45. Hong, S.; Kim, Y. Dynamic Pose Estimation Using Multiple RGB-D Cameras. *Sensors* **2018**, *18*, 3865. [CrossRef]
46. Plagemann, C.; Ganapathi, V.; Koller, D.; Thrun, S. Real-time identification and localization of body parts from depth images. In Proceedings of the 2010 IEEE International Conference on Robotics and Automation, Anchorage, AK, USA, 3–8 May 2010; pp. 3108–3113.

47. Handrich, S.; Al-Hamadi, A.; Lilienblum, E.; Liu, Z. Human bodypart classification using geodesic descriptors and random forests. In Proceedings of the Fifteenth IAPR International Conference on Machine Vision Applications (MVA), Nagoya, Japan, 8–12 May 2017; pp. 318–321.

48. Baak, A.; Müller, M.; Bharaj, G.; Seidel, H.-P.; Theobalt, C. A data-driven approach for real-time full body pose reconstruction from a depth camera. In *Consumer Depth Cameras for Computer Vision*; Springer: London, UK, 2013; pp. 71–98.

49. Mohsin, N.; Payandeh, S. Localization of specific body part by multiple depth sensors network. In Proceedings of the Annual IEEE International Systems Conference (SysCon), Vancouver, BC, Canada, 23–26 April 2018; pp. 1–6.

50. Kong, L.; Yuan, X.; Maharjan, A.M. A hybrid framework for automatic joint detection of human poses in depth frames. *Pattern Recognit.* **2018**, *77*, 216–225. [CrossRef]

51. Carraro, M.; Munaro, M.; Menegatti, E. Skeleton estimation and tracking by means of depth data fusion from depth camera networks. *Robot. Auton. Syst.* **2018**, *110*, 151–159. [CrossRef]

52. Zhang, Y.; Tan, F.; Wang, S.; Yin, B. 3D human body skeleton extraction from consecutive surfaces using a spatial–temporal consistency model. *Vis. Comput.* **2020**, 1–5. [CrossRef]

53. Hu, H.; Li, Z.; Jin, X.; Deng, Z.; Chen, M.; Shen, Y. Curve Skeleton Extraction From 3D Point Clouds Through Hybrid Feature Point Shifting and Clustering. *Comput. Graph. Forum* **2020**, *39*, 111–132. [CrossRef]

54. Sakata, R.; Kobayashi, F.; Nakamoto, H. Development of motion capture system using multiple depth sensors. In Proceedings of the 2017 International Symposium on Micro-NanoMechatronics and Human Science (MHS), Nagoya, Japan, 3–6 December 2017; pp. 1–7.

55. Patil, A.K.; Balasubramanyam, A.; Ryu, J.Y.; Chakravarthi, B.; Chai, Y.H. Fusion of Multiple Lidars and Inertial Sensors for the Real-Time Pose Tracking of Human Motion. *Sensors* **2020**, *20*, 5342. [CrossRef] [PubMed]

56. Zhao, Y.; He, J.; Cheng, H.; Liu, Z. A 2.5 D Thinning Algorithm for Human Skeleton Extraction from a Single Depth Image. In Proceedings of the 2019 Chinese Automation Congress (CAC), Hangzhou, China, 22–24 November 2019; pp. 3330–3335.

57. Xu, T.; An, D.; Wang, Z.; Jiang, S.; Meng, C.; Zhang, Y.; Wang, Q.; Pan, Z.; Yue, Y. 3D Joints Estimation of the Human Body in Single-Frame Point Cloud. *IEEE Access* **2020**, *8*, 178900–178908. [CrossRef]

58. Qi, C.R.; Su, H.; Mo, K.; Guibas, L.J. Pointnet: Deep learning on point sets for 3d classification and segmentation. In Proceedings of the IEEE conference on computer vision and pattern recognition, Honolulu, HI, USA, 21–26 July 2017; pp. 652–660.

59. Zhou, Y.; Tuzel, O. Voxelnet: End-to-end learning for point cloud based 3d object detection. In Proceedings of the IEEE Conference on Computer Vision and Pattern Recognition, Salt Lake City, UT, USA, 18–22 June 2018; pp. 4490–4499.

60. Li, Y.; Bu, R.; Sun, M.; Wu, W.; Di, X.; Chen, B. PointCNN: Convolution on χ-transformed points. In Proceedings of the 32nd International Conference on Neural Information Processing Systems, Montreal, Canada, 2–8 December 2018; pp. 828–838.

61. Wu, W.; Qi, Z.; Fuxin, L. Pointconv: Deep convolutional networks on 3d point clouds. In Proceedings of the IEEE/CVF Conference on Computer Vision and Pattern Recognition, Long Beach, CA, USA, 16–20 June 2019; pp. 9621–9630.

62. Chen, X.; Jiang, K.; Zhu, Y.; Wang, X.; Yun, T. Individual Tree Crown Segmentation Directly from UAV-Borne LiDAR Data Using the PointNet of Deep Learning. *Forests* **2021**, *12*, 131. [CrossRef]

63. Zhang, Z.; Hu, L.; Deng, X.; Xia, S. Weakly Supervised Adversarial Learning for 3D Human Pose Estimation from Point Clouds. *IEEE Trans. Vis. Comput. Graph.* **2020**, *26*, 1851–1859. [CrossRef]

64. Biswas, A.; Admoni, H.; Steinfeld, A. Fast on-board 3D torso pose recovery and forecasting. In Proceedings of the International Conference on Robotics and Automation (ICRA), Montreal, Canada, 20–24 May 2019; pp. 20–24.

65. Özbay, E.; Çınar, A.; Güler, Z. Structured Deep Learning Supported with Point Cloud for 3D Human Pose Estimation. In Proceedings of the 1st International Symposium on Multidisciplinary Studies and Innovative Technologies, Tokat, Turkey, 2–4 November 2017; pp. 304–307.

66. Carraro, M.; Munaro, M.; Burke, J.; Menegatti, E. Real-time marker-less multi-person 3D pose estimation in RGB-Depth camera networks. In Proceedings of the International Conference on Intelligent Autonomous Systems, Porto, Portugal, 13–15 December 2018; pp. 534–545.

67. Schnürer, T.; Fuchs, S.; Eisenbach, M.; Groß, H.-M. Real-time 3D Pose Estimation from Single Depth Images. In Proceedings of the 14th International Joint Conference on Computer Vision, Imaging and Computer Graphics Theory and Applications., Prague, Czech Replublic, 25–27 February 2019; pp. 716–724.

68. Vasileiadis, M.; Bouganis, C.-S.; Tzovaras, D. Multi-person 3D pose estimation from 3D cloud data using 3D convolutional neural networks. *Comput. Vis. Image Underst.* **2019**, *185*, 12–23. [CrossRef]

69. Sengupta, A.; Jin, F.; Zhang, R.; Cao, S. mm-Pose: Real-Time Human Skeletal Posture Estimation using mmWave Radars and CNNs. *IEEE Sens. J.* **2020**, *20*, 10032–10044. [CrossRef]

70. Jiang, Y.; Liu, C.K. Data-driven approach to simulating realistic human joint constraints. In Proceedings of the 2018 IEEE International Conference on Robotics and Automation (ICRA), Prague, Czech Republic, 13–17 August 2018; pp. 1098–1103.

71. Li, S.; Lee, D. Point-to-pose voting based hand pose estimation using residual permutation equivariant layer. In Proceedings of the IEEE Conference on Computer Vision and Pattern Recognition, Long Beach, CA, USA, 16–20 June 2019; pp. 11927–11936.

72. Qi, C.R.; Yi, L.; Su, H.; Guibas, L.J. PointNet++: Deep Hierarchical Feature Learning on Point Sets in a Metric Space. *arXiv* **2017**, arXiv:1706.02413.

73. Chen, Y.; Tu, Z.; Ge, L.; Zhang, D.; Chen, R.; Yuan, J. So-handnet: Self-organizing network for 3d hand pose estimation with semi-supervised learning. In Proceedings of the IEEE International Conference on Computer Vision, Seoul, Korea, 27 October–2 November 2019; pp. 6961–6970.

74. Huang, F.; Zeng, A.; Liu, M.; Qin, J.; Xu, Q. Structure-aware 3d hourglass network for hand pose estimation from single depth image. *arXiv* **2018**, arXiv:1812.10320.

75. Ge, L.; Ren, Z.; Yuan, J. Point-to-point regression pointnet for 3d hand pose estimation. In Proceedings of the European conference on computer vision (ECCV), Munich, Germany, 8–14 September 2018; pp. 475–491.

76. Ge, L.; Cai, Y.; Weng, J.; Yuan, J. Hand pointnet: 3d hand pose estimation using point sets. In Proceedings of the IEEE Conference on Computer Vision and Pattern Recognition, Salt Lake City, UT, USA, 18–22 June 2018; pp. 8417–8426.

77. Reale, M.J.; Klinghoffer, B.; Church, M.; Szmurlo, H.; Yin, L. Facial Action Unit Analysis through 3D Point Cloud Neural Networks. In Proceedings of the 14th IEEE International Conference on Automatic Face & Gesture Recognition (FG 2019), Lille, France, 14–18 May 2019; pp. 1–8.

78. Ge, L.; Liang, H.; Yuan, J.; Thalmann, D. 3d convolutional neural networks for efficient and robust hand pose estimation from single depth images. In Proceedings of the IEEE Conference on Computer Vision and Pattern Recognition, Honolulu, HI, USA, 21–26 July 2017; pp. 1991–2000.

79. Bekhtaoui, W.; Sa, R.; Teixeira, B.; Singh, V.; Kirchberg, K.; Chang, Y.-J.; Kapoor, A. View Invariant Human Body Detection and Pose Estimation from Multiple Depth Sensors. *arXiv* **2020**, arXiv:2005.04258.

80. van Sabben, D.; Ruiz-Hidalgo, J.; Cuadros, X.S.; Casas, J.R. Collaborative voting of 3D features for robust gesture estimation. In Proceedings of the 2017 IEEE International Conference on Acoustics, Speech and Signal Processing (ICASSP), New Orleans, LA, USA, 5–9 March 2017; pp. 1677–1681.

81. Xia, S.; Zhang, Z.; Su, L. Cascaded 3d full-body pose regression from single depth image at 100 fps. In Proceedings of the 2018 IEEE Conference on Virtual Reality and 3D User Interfaces (VR), Tuebingen/Reutlingen, Germany, 18–22 March 2018; pp. 431–438.

82. Tsai, M.-H.; Chen, K.-H.; Lin, I.-C. Real-time upper body pose estimation from depth images. In Proceedings of the 2015 IEEE International Conference on Image Processing (ICIP), Québec City, QC, Canada, 27–30 September 2015; pp. 2234–2238.

83. Dinh, D.-L.; Han, H.-S.; Jeon, H.J.; Lee, S.; Kim, T.-S. Principal direction analysis-based real-time 3D human pose reconstruction from a single depth image. In Proceedings of the 4th International Symposium on Information and Communication Technology, Da Nang, Vietnam, 5–6 December 2013; pp. 206–212.

84. Park, S.; Yong Chang, J.; Jeong, H.; Lee, J.-H.; Park, J.-Y. Accurate and efficient 3d human pose estimation algorithm using single depth images for pose analysis in golf. In Proceedings of the IEEE Conference on Computer Vision and Pattern Recognition Workshops, Honolulu, HI, USA, 21–26 July 2017; pp. 49–57.

85. Kim, J.; Kim, H. Robust geodesic skeleton estimation from body single depth. In Proceedings of the International Conference on Advanced Concepts for Intelligent Vision Systems, Poitiers, France, 24–27 September 2018; pp. 342–353.

86. Huang, C.-H.; Boyer, E.; do Canto Angonese, B.; Navab, N.; Ilic, S. Toward user-specific tracking by detection of human shapes in multi-cameras. In Proceedings of the IEEE Conference on Computer Vision and Pattern Recognition, Boston, MA, USA, 7–12 June 2015; pp. 4027–4035.

87. Handrich, S.; Waxweiler, P.; Werner, P.; Al-Hamadi, A. 3D Human Pose Estimation Using Stochastic Optimization in Real Time. In Proceedings of the 25th IEEE International Conference on Image Processing (ICIP), Athens, Greece, 7–10 October 2018; pp. 555–559.

88. Yub Jung, H.; Lee, S.; Seok Heo, Y.; Dong Yun, I. Random tree walk toward instantaneous 3d human pose estimation. In Proceedings of the IEEE Conference on Computer Vision and Pattern Recognition, Boston, MA, USA, 7–12 June 2015; pp. 2467–2474.

89. Ganapathi, V.; Plagemann, C.; Koller, D.; Thrun, S. Real-time human pose tracking from range data. In Proceedings of the European conference on computer vision, Firenze, Italy, 7–13 October 2012; pp. 738–751.

90. Ganapathi, V.; Plagemann, C.; Koller, D.; Thrun, S. Real time motion capture using a single time-of-flight camera. In Proceedings of the 2010 IEEE Computer Society Conference on Computer Vision and Pattern Recognition, San Francisco, CA, USA, 13–18 June 2010; pp. 755–762.

91. Li, W.; Zhang, Z.; Liu, Z. Action recognition based on a bag of 3d points. In Proceedings of the 2010 IEEE Computer Society Conference on Computer Vision and Pattern Recognition-Workshops, San Francisco, CA, USA, 13–18 June 2010; pp. 9–14.

92. Shahroudy, A.; Liu, J.; Ng, T.-T.; Wang, G. Ntu rgb+ d: A large scale dataset for 3d human activity analysis. In Proceedings of the IEEE conference on computer vision and pattern recognition, Las Vegas, NV, USA, 27–30 June 2016; pp. 1010–1019.

93. Nguyen, T.-N.; Meunier, J. *Walking Gait Dataset: Point Clouds, Skeletons and Silhouettes*; Technical Report No. 1379; University of Montreal: Montreal, QC, Canada, 2018.

94. Ofli, F.; Chaudhry, R.; Kurillo, G.; Vidal, R.; Bajcsy, R. Berkeley mhad: A comprehensive multimodal human action database. In Proceedings of the 2013 IEEE Workshop on Applications of Computer Vision (WACV), Clearwater Beach, FL, USA, 15–17 January 2013; pp. 53–60.

95. Bloom, V.; Makris, D.; Argyriou, V. G3D: A gaming action dataset and real time action recognition evaluation framework. In Proceedings of the 2012 IEEE Computer Society Conference on Computer Vision and Pattern Recognition Workshops, Providence, RI, USA, 16–21 June 2012; pp. 7–12.

96. Bloom, V.; Argyriou, V.; Makris, D. G3di: A gaming interaction dataset with a real time detection and evaluation framework. In Proceedings of the European Conference on Computer Vision, Zurich, Switzerland, 6–12 September 2014; pp. 698–712.

97. Yun, K.; Honorio, J.; Chattopadhyay, D.; Berg, T.L.; Samaras, D. Two-person interaction detection using body-pose features and multiple instance learning. In Proceedings of the 2012 IEEE Computer Society Conference on Computer Vision and Pattern Recognition Workshops, Providence, RI, USA, 16–21 June 2012; pp. 28–35.

98. Holt, B.; Ong, E.-J.; Cooper, H.; Bowden, R. Putting the pieces together: Connected poselets for human pose estimation. In Proceedings of the 2011 IEEE international conference on computer vision workshops (ICCV workshops), Barcelona, Spain, 6–13 November 2011; pp. 1196–1201.

99. CMU Mocap Database. Available online: http://mocap.cs.cmu.edu (accessed on 11 December 2012).

100. Yang, J.; Franco, J.-S.; Hétroy-Wheeler, F.; Wuhrer, S. Estimation of human body shape in motion with wide clothing. In Proceedings of the European Conference on Computer Vision, Amsterdam, The Netherlands, 11–14 October 2016; pp. 439–454.

101. Saint, A.; Shabayek, A.E.R.; Aouada, D.; Ottersten, B.; Cherenkova, K.; Gusev, G. Towards Automatic Human Body Model Fitting to a 3D Scan. In Proceedings of the 8th International Conference and Exhibition on 3D Body Scanning and Processing Technologies, Montreal, QC, Canada, 11–12 October 2017; pp. 274–280.

102. Mishra, G.; Saini, S.; Varanasi, K.; Narayanan, P. Human Shape Capture and Tracking at Home. In Proceedings of the 2018 IEEE Winter Conference on Applications of Computer Vision (WACV), Lake Tahoe, NV, USA, 12–15 March 2018; pp. 390–399.

103. Wang, J.; Lu, Z.; Liao, Q. Estimating Human Shape Under Clothing from Single Frontal View Point Cloud of a Dressed Human. In Proceedings of the 2019 IEEE International Conference on Image Processing (ICIP), Taipei, Taiwan, 22–25 September 2019; pp. 330–334.

104. Yao, L.; Peng, X.; Guo, Y.; Ni, H.; Wan, Y.; Yan, C. A Data-Driven Approach for 3D Human Body Pose Reconstruction from a Kinect Sensor. *JPhCS* **2018**, *1098*, 012024. [CrossRef]

105. Kim, W.; Ramanagopal, M.S.; Barto, C.; Yu, M.-Y.; Rosaen, K.; Goumas, N.; Vasudevan, R.; Johnson-Roberson, M. PedX: Benchmark dataset for metric 3-D pose estimation of pedestrians in complex urban intersections. *IEEE Robot. Autom. Lett.* **2019**, *4*, 1940–1947. [CrossRef]

106. Du, Y.; Wang, W.; Wang, L. Hierarchical recurrent neural network for skeleton based action recognition. In Proceedings of the IEEE conference on computer vision and pattern recognition, Boston, MA, USA, 7–12 June 2015; pp. 1110–1118.

107. Wang, J.; Liu, Z.; Wu, Y.; Yuan, J. Mining actionlet ensemble for action recognition with depth cameras. In Proceedings of the 2012 IEEE Conference on Computer Vision and Pattern Recognition, Providence, RI, USA, 18–20 June 2012; pp. 1290–1297.

108. Wang, J.; Liu, Z.; Wu, Y.; Yuan, J. Learning actionlet ensemble for 3D human action recognition. *IEEE Trans. Pattern Anal. Mach. Intell.* **2013**, *36*, 914–927. [CrossRef]

109. Khan, M.S.; Ware, A.; Karim, M.; Bahoo, N.; Khalid, M.J. Skeleton based Human Action Recognition using a Structured-Tree Neural Network. *Eur. J. Eng. Technol. Res.* **2020**, *5*, 849–854. [CrossRef]

110. Zhang, E.; Chen, W.; Zhang, Z.; Zhang, Y. Local surface geometric feature for 3D human action recognition. *Neurocomputing* **2016**, *208*, 281–289. [CrossRef]

111. Khokhlova, M.; Migniot, C.; Dipanda, A. 3D Point Cloud Descriptor for Posture Recognition. In Proceedings of the Computer Vision, Imaging and Computer Graphics Theory and Applications - 13th International Joint Conference, Funchal, Madeira, Portugal, 27–29 January 2018; pp. 161–168.

112. Liu, M.; Liu, H.; Chen, C. Enhanced skeleton visualization for view invariant human action recognition. *Pattern Recognit.* **2017**, *68*, 346–362. [CrossRef]

113. Wang, S.; Zuo, X.; Wang, R.; Cheng, F.; Yang, R. A generative human-robot motion retargeting approach using a single depth sensor. In Proceedings of the 2017 IEEE International Conference on Robotics and Automation (ICRA), Singapore, 29 May–3 June 2017; pp. 5369–5376.

114. Wang, S.; Zuo, X.; Wang, R.; Yang, R. A Generative Human-Robot Motion Retargeting Approach Using a Single RGBD Sensor. *IEEE Access* **2019**, *7*, 51499–51512. [CrossRef]

115. Kostavelis, I.; Vasileiadis, M.; Skartados, E.; Kargakos, A.; Giakoumis, D.; Bouganis, C.-S.; Tzovaras, D. Understanding of human behavior with a robotic agent through daily activity analysis. *Int. J. Soc. Robot.* **2019**, *11*, 437–462. [CrossRef]

116. Kim, Y. Dance motion capture and composition using multiple RGB and depth sensors. *Int. J. Distrib. Sens. Netw.* **2017**, *13*, 155014771769608. [CrossRef]

117. Wang, H.; Liang, W.; Yu, L.-F. Transferring objects: Joint inference of container and human pose. In Proceedings of the IEEE International Conference on Computer Vision, Venice, Italy, 22–29 October 2017; pp. 2933–2941.

118. Patruno, C.; Marani, R.; Cicirelli, G.; Stella, E.; D'Orazio, T. People re-identification using skeleton standard posture and color descriptors from RGB-D data. *Pattern Recognit.* **2019**, *89*, 77–90. [CrossRef]

119. Fu, T.; Chaine, R.; Digne, J. FAKIR: An algorithm for revealing the anatomy and pose of statues from raw point sets. *Computer Graphics Forum* **2020**, *39*, 375–385. [CrossRef]

120. Varadarajan, S.; Tiwari, N.; Datta, P.; Silva, A.P.M.; Tickoo, O.; Carroll, E. Age classification of humans based on image depth and human pose. U.S. Patent 10,540,545, 21 January 2020.

121. Desai, K.; Prabhakaran, B.; Raghuraman, S. Combining skeletal poses for 3D human model generation using multiple Kinects. In Proceedings of the 9th ACM Multimedia Systems Conference, Amsterdam, The Netherlands, 12–15 June 2018; pp. 40–51.

122. Jatesiktat, P.; Anopas, D.; Ang, W.T. Personalized markerless upper-body tracking with a depth camera and wrist-worn inertial measurement units. In Proceedings of the 40th Annual International Conference of the IEEE Engineering in Medicine and Biology Society (EMBC), Honolulu, HI, USA, 18–21 July 2018; pp. 1–6.

MDPI
St. Alban-Anlage 66
4052 Basel
Switzerland
Tel. +41 61 683 77 34
Fax +41 61 302 89 18
www.mdpi.com

Sensors Editorial Office
E-mail: sensors@mdpi.com
www.mdpi.com/journal/sensors

www.ingramcontent.com/pod-product-compliance
Lightning Source LLC
LaVergne TN
LVHW070843170726
843515LV00005B/559